STO

ACPL ITEM DISCARDED

621.3897 M56a
Mether, Calvin E.
Audiovisual fundamentals

Y0-AAD-617

Audiovisual Fundamentals

DO NOT REMOVE
CARDS FROM POCKET

ALLEN COUNTY PUBLIC LIBRARY

FORT WAYNE, INDIANA 46802

You may return this book to any agency, branch,
or bookmobile of the Allen County Public Library.

DEMCO

Audiovisual Fundamentals
**Basic Equipment Operation
Simple Materials Production**

Fourth Edition

Calvin E. Mether
University of Iowa

John R. Bullard
University of Iowa (retired)

Bill Martin
University of Iowa

wcb

Wm. C. Brown Publishers
Dubuque, Iowa

Pages 14; 16; 18; 20; 50; 52; 54; 56; 88; 90; 92; 94; 96; 98; 100

The illustrated projectors were manufactured by Eastman Kodak Company.

Cover credit copyright © Paul Light/Light Wave

Copyright © 1974, 1979, 1984, 1989 by Wm. C. Brown Publishers. All rights reserved

Library of Congress Catalog Card Number: 88-71908

ISBN 0-697-06133-7

No part of this publication may be reproduced, stored in a retrieval system, or transmitted, in any form or by any means, electronic, mechanical, photocopying, recording, or otherwise, without the prior written permission of the publisher.

Printed in the United States of America by Wm. C. Brown Publishers 2460 Kerper Boulevard, Dubuque, IA 52001

10 9 8 7 6 5 4 3 2 1

Contents

Preface	vii
Acknowledgments	ix
Supplemental Audiovisual Materials	xi
How to Use This Book	xiii
General Equipment Operation Principles	xv
I. Projection Principles	1
Overhead Projector (Beseler)	5
Opaque Projector (Beseler)	9
Multiple-load Slide Projector (Kodak)	15
Portable Screen (Da-Lite)	23
II. Audio Principles	28
Record Player (Newcomb)	31
Audiocassette Recorder with Slide/ Sync System (Sharp)	37
III. AV Equipment Principles	46
Preparation of a Synchronized Slide/ Tape Program (Wollensak)	48
Coyote, Tascam 225 Syncaset, and AVL Synchronizer	59
Slot-load 16mm Sound Motion Picture Projector (Eiki)	79
Manual-thread 16mm Sound Motion Picture Projector (Kodak)	89
Filmstrip Projector with Audiocassette Playback (DuKane Micromatic)	103
Commonly Used Cable Plugs and Jacks	108
IV. Video Systems	111
Studio Production	111
Videocassette Recorder/Player (Sony)	113
Electronic Field Production	131
1/2" VHS Camcorder Recorder/Player (Panasonic)	133
V. Troubleshooting Guide for Instructional Equipment	144
VI. Microcomputers as Audiovisual Tools	154
Overview	154
Software	154
Hardware	155
Using a Microcomputer as a Projection Device	157
Using a Microcomputer to Prepare Classroom Materials	157
Using a Microcomputer as an Instructional Tool	159
VII. Photographic Copystand Project	163
VIII. Lettering Principles	170
Dry Transfer	172
Cardboard Stencils	173
IX. Thermal Copying	174
Transparencies	174
Thermal Spirit Master Preparation	177
X. Spirit Duplication	179
Quantity Duplication	179
Direct Spirit Master Preparation	181
XI. Materials Preservation	183
Dry Mounting	183
Laminating	185
XII. The Copyright Law and Its Implications	187
Appendix	196
Study Questions	197
Projection (General)	197
Overhead Projector	199
Opaque Projector	200
Filmstrip Projector with Audiocassette Playback	201
Multiple-load Slide Projector	202
Audio (General)	203
Audiocassette Tape Recorder	204
Coyote, Tascam 225, and 3 Projectors	206
Slot-load 16mm Sound Motion Picture Projector	207
Manual-thread 16mm Sound Motion Picture Projector	208
Videotape Systems	209
Microcomputers	211
Photographic Copystand	212
Lettering	213
Thermal Copying	214

    Spirit Duplication   215
    Materials Preservation   216
    The Copyright Law and its Implications   217
Study Question Answer Sheet   218
Materials Production Assignment Sheets   220
    1. Dry-Transfer Lettering   221
    2. Cardboard Stencil Lettering   222
    3. Thermal Transparency   223
    4. Thermal Spirit Master   224
    5. Direct Spirit Master   225
    6. Dry Mounting   226
    7. Laminating (With the Dry-mount Press)   227
Instructional Equipment Checkout Sheet   228

Preface

The ability to operate basic audiovisual equipment and produce simple instructional materials is a prerequisite to the effective use of media. Lack of confidence on the part of a potential media-user frequently means that presentations are handled in less effective conventional ways.

Similarly, most of us have seen or heard presentations where the use of media was so ineffective (e.g., garbled audio, unreadable visuals, out-of-focus images, distracting extraneous light and noise) that a straight lecture-type presentation would have been better.

This book was developed to help us provide several hundred pre-service teachers each semester with appropriate hands-on experience with audiovisual equipment. The self-instruction format enables our limited staff to put its energies into evaluation of student products and operational skills. It allows students to practice and work in the lab at any schedule convenient to them.

Preliminary forms of this book have been used at our institution since the Fall semester of 1972. Hundreds of students have learned how to operate the pieces of audiovisual presentation and production equipment that are explained in the book.

We have used simple line drawings because photographs typically include too much information, i.e., important features are often hard to locate, and unimportant elements distract the learner. Insignificant distractors (nameplates, texture variation, screws, etc.) have been eliminated from the drawings when it is practical.

An instructor's manual and course syllabus are available to instructors who request them. These provide an explanation of the scheduling, grading, and checkout procedures that have enabled us to individually grade large numbers of students with limited staff and facilities.

While we have attempted to provide generalized instructions about each type of equipment, we have given detailed instructions about specific models. We have selected reasonably complex and popular models in the belief that the major hurdle to operating a piece of audiovisual equipment is overcoming the fear of complex electronic devices and developing some confidence in the ability to operate them. A person who has become proficient in using one model of 16mm projector or VCR, for example, is usually not reluctant to spend a few minutes figuring out how to operate another model.

There has been no attempt to cover all types of audiovisual machines. We have selected those that we felt a classroom teacher was most likely to have the opportunity to use. Some, such as portable audiocassette recorders, are so simple to operate, that we have not included them on the assumption that students would be able to operate them without any problems if they can operate a 16mm motion picture projector.

We have not attempted to be exhaustive in our discussion of each piece of equipment or process. We have not referred to parts that are not used during the operations required.

The list of Essential Parts at the beginning of each unit includes those items that the learner must "do something with" during the learning process. This list is not intended to imply that these are the basic elements of the piece of equipment; e.g., failure to call out lamps, motors, and fans does not imply that these are not essential parts.

No attempt has been made to cover mixing, patching, or editing of various media, or to discuss problems that can best be handled by a competent Audiovisual Specialist.

We should emphasize that this book was designed to teach how to operate basic equipment and produce simple materials. We have not attempted to discuss when, why, and under what conditions you should use any of the various types of educational resources. Selection and use of these media are important and difficult tasks that are not within the scope of this book or of the one-semester hour course it was designed to serve.

The fourth edition of *Audiovisual Fundamentals* has been revised to include equipment and procedures that are in common use today. A slot-load 16mm motion picture projector has been added, and the VHS camcorder is presented as a convenient, mobile production alternative where motion and audio are needed. Photographic copystand equipment and techniques are included for preparation of visuals. Production of a slide/tape synched presentation is illustrated and techniques are noted. Computers are presented in greater depth as versatile pieces of production and presentation equipment. A section is also included on interpretation of copyright provisions as they apply to AV production and presentation.

C. E. M.

W. E. M.

J. R. B.

ACKNOWLEDGMENTS

Revision Coordinator

Jamie Achrazoglou, University of Iowa

Contributing Authors

John Achrazoglou, Coordinator, Computer Resources Lab, College of Education, University of Iowa

Kathleen Tessmer, Assistant Professor, School of Library and Information Science, College of Education, Division of Psychological and Quantitative Foundations, University of Iowa

Doris Hildreth, Instructional Design, College of Education, University of Iowa

Cynthia Lake, Instructional Design, College of Education, University of Iowa

Barbara McElroy Knight, Instructional Design, College of Education, University of Iowa

Kressa Jane Peck, Instructional Design, College of Education, University of Iowa

Design and Evaluation

Scott Popham, Secretary, AV Production Lab, College of Education, University of Iowa

Juanita Arellanos, Graduate Assistant, College of Education, University of Iowa

Ron Osgood, Indiana University

Paul Herrin, Graduate Student, College of Education, University of Iowa

Tim Kane, Graduate Assistant, College of Education, University of Iowa

Debra Hohn, University of Iowa

Byron Morales, University of Iowa

Dennis Weir, University of Iowa

Aimee Sturm, University of Iowa

Illustration models courtesy of:

- Audio Visual Labs
- Charles Beseler Company
- Da-Lite Corporation
- Dukane Corporation
- Eiki International Inc.
- TEAC Corporation of America
- Eastman Kodak Company
- Panasonic Corporation
- Seal Corporation
- Sony Corporation of America
- Sharp Corporation
- IBM Corporation
- Mindscape, Inc.
- Apple Computer, Inc.

Supplemental Audiovisual Materials

A series of training color sound filmstrips is available showing the simple techniques involved in dry mounting to prepare visuals for the classroom, library, or home as well as for business or sales presentations, lectures, and displays. They are available from International Film Bureau Inc., 332 South Michigan Avenue, Chicago, Illinois 60604.

Graphic Production Techniques—Series I.

Dry Mounting Flat Materials
Cloth Backing and Hinging
Lamination and Picture Transfer
Press Maintenance and Care

Graphic Production Techniques—Series II.

Dry Mounting RC Prints
Creating Display Materials
Matte Framing and Some Other Ideas
Special Applications

A series of training videotapes (listed in the right-hand column) on commonly used audiovisual equipment and processes has been developed for use with this text. They are intended for classroom instruction, in-service training, workshops, training equipment operators, and for self-instruction. For information about the tapes, please contact William E. Martin, N157A, Lindquist Center, College of Education, University of Iowa, Iowa City, Iowa 52242.

Titles

1. Beseler Opaque Projector
2. Draper Portable Tripod Screen
3. Panasonic Portable Audiocassette Recorder
4. 3M Overhead Projector
5. Kodak Ektagraphic III 35mm Slide Projector
6. Dukane Sound Filmstrip Projector
7. Kodak Pageant 16mm Sound Projector
8. Sony Single Camera VCR System (black and white)
9. 3M Thermofax
10. A. B. Dick Spirit Duplicator
11. Basic Lettering for Audiovisual Materials
12. Dry Mounting Audiovisual Materials
13. Laminating Audiovisual Materials
14. General Operating Principles for AV Equipment
15. Photographic Copystand
16. Eiki Slot-Load 16mm Sound Projector
17. Production of a Slide/Tape Program (Wollensak)
18. Script Writing
19. Basic Audio Production
20. Basic Lighting for Video
21. Basic Video Camera Techniques
22. Panasonic Cam/Corder
23. Basic Videotape Editing
24. Coyote, Tascam 225 Syncaset, and AVL Synchronizer
25. Sharp Audio Cassette Recorder with Slide/Sync System
26. Sony Single Camera/VCR Monitor System (color)

How to Use This Book

This book was designed to provide a means for learners to develop confidence and skill in using selected instructional resources in the absence of an instructor. Experience with preliminary drafts of this book, and hundreds of students, has demonstrated its effectiveness as a device for self-instruction.

No prior experience with audiovisual equipment or the production of instructional materials is expected, but access to, and actual practice with the equipment described is a necessary part of the learning process.

The operation of basic audiovisual presentation equipment is covered in the first four chapters, and the production of simple instructional materials is explained in the last seven chapters. The introductions to the chapters provide a discussion of the general principles that apply to the processes explained within.

The step-by-step instructions on how to operate presentation equipment are provided in a three-column format with a single reference illustration. The left-hand column indicates general procedures that apply to most models, the center column provides instructions specific to the model depicted, and the right-hand column attempts to supplement and clarify the process. To adequately understand and carry out the instructions, it is imperative that all three columns are carefully read during the initial learning experience.

Because control mechanisms (levers, switches, buttons, etc.) and other operational parts vary in conformation and location on different brands and models, general terms (power control) rather than descriptive ones (on-off switch) for the specific model depicted have been used. Noteworthy variations on specific models will usually be mentioned in the third column. The absence of functional controls similar to those provided on the model depicted, or the presence of additional controls, will generally be indicated on the first page or in the copy of each unit.

Some of the equipment can be mastered by referring to the book during the first attempt, then closing the book and repeating the process. The more complex processes should be practiced with progressively fewer references to the book until the user can accomplish the task efficiently and confidently without referring to the instructions. With practice, the learner will become more efficient and more confident. Hands-on practice is a necessary part of developing any psychomotor skills — they cannot be learned by just reading about them.

A Troubleshooting Guide for Instructional Equipment is provided at the end of the section to assist in diagnosing problems not discussed in the text.

The section that provides visualized instructions on how to produce simple instructional materials is followed by Assignment Sheets, which specify simple exercises to provide experience with the process and provide evidence of ability to complete an acceptable product.

Experience has shown that this section can cause the learner some frustration. Several factors seem to be operating here. The most common problem is that learners prefer to be shown how to do something rather than learn about it on their own. If one student appears to be having success, there is a tendency to ask rather than read. This tendency can cause unfortunate results and wasted materials. Most of the problems could be avoided if users would read the step-by-step instructions carefully. A second problem relates to materials spoilage. The anxiety level of students increases when things do not go right because spoiled materials are expensive. A certain percentage of failure can be anticipated even if one has considerable experience with the process. Finally, it must be recognized that the materials or the equipment may be at fault. If instructions have been carefully followed and results are unsatisfactory, the learner should seek help from a qualified advisor.

General Equipment Operation Principles

Audiovisual equipment is designed and produced to varying degrees of precision, and to meet a specific need. Before using equipment for purposes for which it was not designed, careful consideration should be given to the possible results.

It is unwise to force any controls; if a piece of audiovisual equipment does not operate properly, check set-up procedures. The troubleshooting section in this book may give some clue to the cause of the problem. After you have determined that the difficulty is beyond your capability, report it to the individual who is responsible for the equipment.

Normal use may cause knobs, levers, or adjusting screws to work loose. These may be finger-tightened by the operator and reported.

Dust, raised by custodial functions, can settle on surfaces upon which film or tape travels. The dust or dirt is picked up by the film or tape and carried to an aperture or gate where it piles up and will cause scratches and unnecessary wear. It is advisable to cover all pieces of audiovisual equipment when they are not in use and to have them cleaned periodically.

Make a habit of looking at all power cords to determine whether the ends are in dangerous condition before using. Stray wires, loose ends, or broken insulation can cause a short circuit, an electrical shock , or a fire.

Always remove power cords by grasping the plug rather than pulling the cord. Newer machines that have a grounded three-wire plug may require an adapter.

Emergencies may demand simple adjustments or minor repairs by the operator. Competent technicians should perform major repairs or adjustments that require specialized training, equipment, or facilities.

Projectors of various types, record players, and recorders generate heat while in operation. It is vital that provision be made for the free circulation of air. The entire cord should be removed from the power cord storage compartment to allow for adequate air circulation. Most machines have fans that operate when the equipment is in operation and some have thermostatic controls that cause them to circulate air until the heat has been lowered to safe levels. Never unplug a machine after turning the projection lamp off if a thermostatically controlled fan continues to run.

When disconnecting cables, do not use force. Some cables require the operator to squeeze release buttons on the cable ends. If a projection lens resists removal or focusing, it may be necessary to loosen a set screw or realign the mechanism.

Controls on all machines should be placed in "neutral" or "off" for storage to avoid damage to internal drive mechanisms.

Chapter I Projection Principles

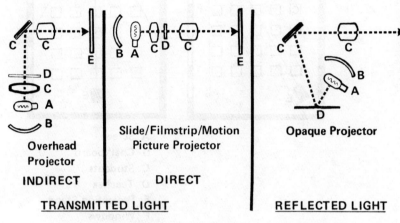

BASIC PROJECTION SYSTEM COMPONENTS

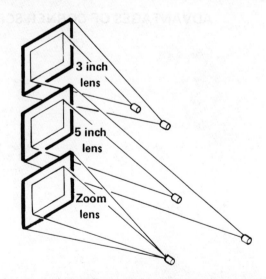

EFFECT OF PROJECTION LENS LENGTH AND PROJECTOR-TO-SCREEN DISTANCE

The basic components necessary for a projection system are:

A. source of light
B. reflector to control the direction of the light
C. lens system (which may include mirrors)
D. means for supporting material to be projected
E. surface for viewing the projected image

Many projection systems now use a bulb with a built-in reflector to produce transmitted light of high intensity with relatively low wattage.

Projection lenses are available in a variety of focal lengths. Selection is affected by type of projector used and projection distance. In many instructional situations, there are no lens choices so the operator must adjust image size on the screen by moving the projector farther from the screen to make the image size larger and closer to decrease its size. Image brightness is affected by the distance from projector to screen, because the farther light travels, the less brilliant is the projected image on the screen. Under some conditions, it may be neither possible nor desirable to fill the screen area. A brighter image may be obtained by a short projection distance that would not fill the screen. Zoom lenses permit the operator to change image size from a given location to more completely fill the screen area. The focal length is changed by rotating of the lens barrel and then refocusing the screen image.

Prefocusing enables the operator to determine best projection size and alignment.

Projected images may be displayed by:

A. front projection in which the projector is opposite a reflective screen surface
B. rear projection in which the projector is behind a translucent screen surface

Mirrors are frequently used in rear projection systems to shorten the distance from projector to screen and to facilitate equipment placement.

A glass-beaded screen is often used in classrooms where the projection distance is greater than the room width. The brilliance of the reflected image is reduced for viewers seated at angles greater than twenty-five degrees on either side of the projection axis. Matte-surfaced screens are a better choice for rooms where room width and projection are nearly equal because the optimum viewing angle is broader. Lenticular screens combine the advantages of both beaded and matte screens in that they serve wider groups than beaded and more rows than matte screens.

Advantages of placing a screen in the corner of a room on a window wall include:

A. an unobstructed view for more students (especially with the overhead projector)
B. a position where less ambient light from windows will fall on the screen resulting in a brighter image
C. unobstructed use of the chalkboard
D. optimum viewing angle for more students

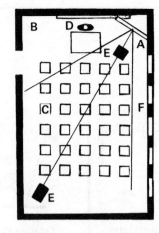
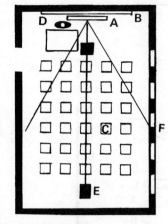

A Screen
B Chalkboard
C Students
D Teacher
E Projection Stand
F Windows

ADVANTAGES OF CORNER SCREEN

For optimal viewing, observer-to-screen distance should not be less than two times nor more than six times the width of the projected image. If possible, the lower edge of the screen should be high enough to be easily seen by the back-row viewers.

A "keystone effect" is caused by projected light rays falling on one part of the screen having to travel a greater distance than those falling on another part of the screen. It can be eliminated by adjusting the screen so that its surface is perpendicular to the projection axis (typically, tilt the top of the screen forward and/or adjust the floor position of the projector relative to the screen).

Lamp life is relatively short and projection lamps are expensive. They range from $8 to $40 for a single lamp. When a projection lamp is hot, it is most vulnerable to breakage. Let the lamp cool slowly and naturally before moving the machine. Condensing lenses and heat-absorbing filters can be cracked or distorted by sudden exposure to cold air.

Lenses and film gates should be cleaned regularly for most effective projection. Because film moves through a projector top to bottom, dirt and emulsion will collect on the lower edges of the pressure plate and the aperture plate. Use a camel's-hair brush, lens cleaning fluid, and lens tissue when cleaning optical systems. Cotton swabs are also useful for cleaning areas that contact film surfaces.

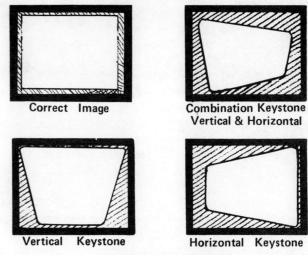

KEYSTONE EFFECT

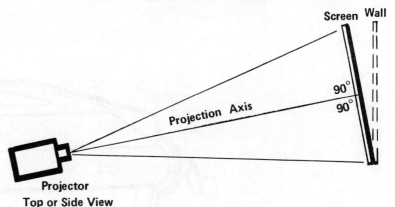

CAUSE AND CURE OF KEYSTONE EFFECT

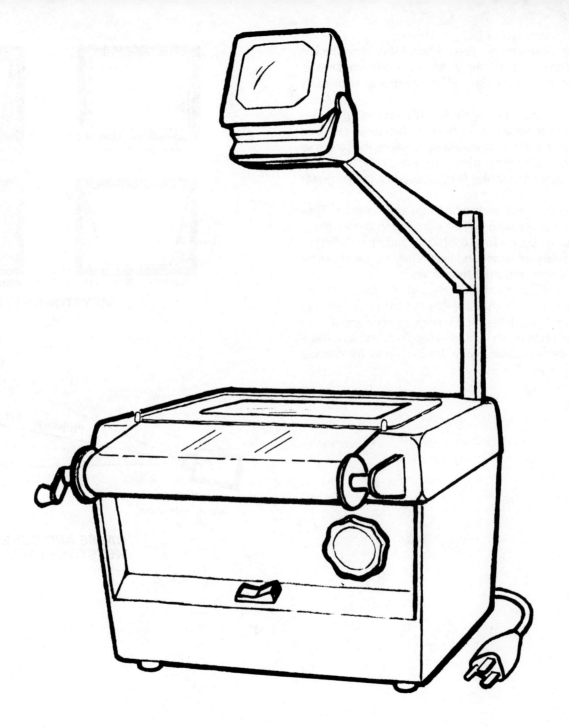

Overhead Projector

The overhead projector is used at the front of the room with the image projected onto a screen behind the user who faces the audience. Room darkening is not required due to the quantity of light that is transmitted through the large transparencies.

Materials must be transparent in order to be projected with the overhead projector. Commercially prepared or handmade transparencies are placed on the stage so that the user can read the material while facing the audience. The large size transparencies allow the user to "write on," "color in," and "point at" specific details while discussing them. Silhouettes, plastic tools, and transparent materials for science experiments are only a few of the possible software items that can be used on the overhead projector.

Objectives

Observable Behavior

1. Project a transparency on a screen.

2. Restore projector to storage conformation.

Conditions

Time limit 1 minute.

Time limit 1 minute.

Level of Acceptable Performance

The projected image must be right side up, in focus, maximum size with minimum keystoning.

Power cord neatly wound.

Essential Parts

Typical

Power cord
Stage
Upper lens assembly
Controls:
    Elevation
    Focus
    Power

Additional on Beseler

Acetate supply rolls
Registration guides

Available on Some Models

Hinged shelves for transparency support
Storage compartments for transparencies
Mechanisms to create the illusion of motion
Fan control

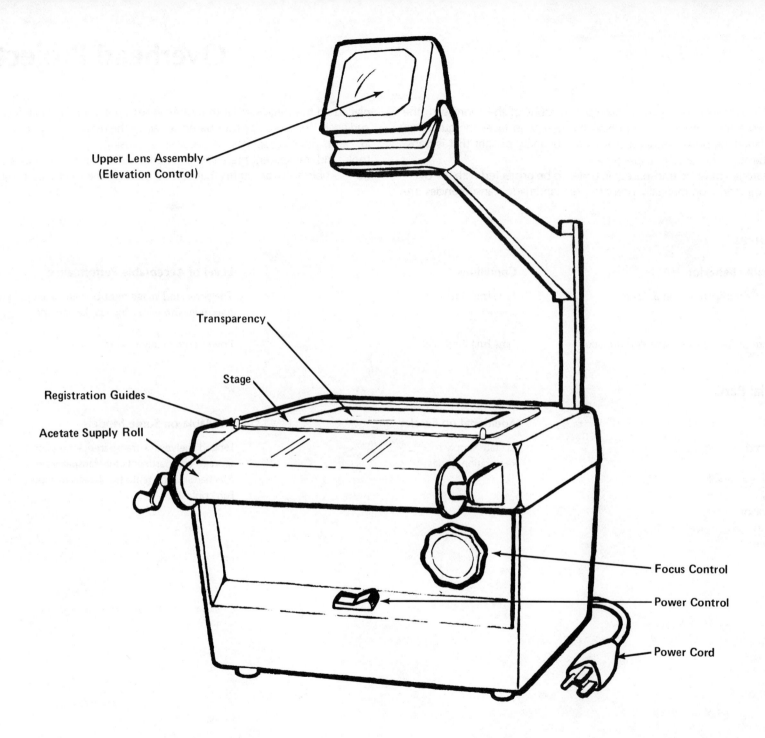

General Instructions

1. Read Projection Principles before proceeding.

2. Tie *power cord* around bottom of a projection stand leg before connecting it to a suitable power source or wall outlet.

3. Turn lamp **on** with *power control*.

4. Position *transparency* on *stage* of projector.

5. Adjust position of projector and elevation of image to fill screen.

6. Adjust *focus control*.

7. Turn lamp **off** with *power control*.

8. Disconnect *power cord* and restore projector to storage conformation.

Beseler Overhead Instructions

Power cord is wound around bracket on back of projector.

Upper lens assembly (*elevation control*) can be tilted to elevate the image.

Supplementary Information

The *power cord* may be stored inside an access door on the machine. The projector will not operate until the door is tightly closed.

Aim projector so that light falls on screen over operator's shoulder.

Transparency should be right side up so that operator can read it while facing audience.

It is desirable to mount a transparency on a cardboard frame that covers the entire stage area to prevent distracting light around the border. It will also keep the transparency flat, provide a place for lecture notes, and facilitate handling and storage. Some projectors have *registration* guides to insure correct positioning of subsequent transparencies. Rolls of acetate are provided on many machines to enable user to write on the *stage* with a marking pencil or pen. Do not write directly on glass *stage*.

See information on keystoning.

If it is desirable to gain the students' attention by terminating a visual image projected from an overhead projector, a file folder or sheet of tagboard (10 × 10 inches) will effectively block the light path when placed on the stage of the machine.

Typically, the cooling fan is thermostatically controlled and should be left running until projector has cooled.

The projector should be carried by transport handle if it has one, or with both hands under the lower sides of the projector. Never carry an overhead projector by the *upper lens assembly*, supporting post, or sides of the *stage*.

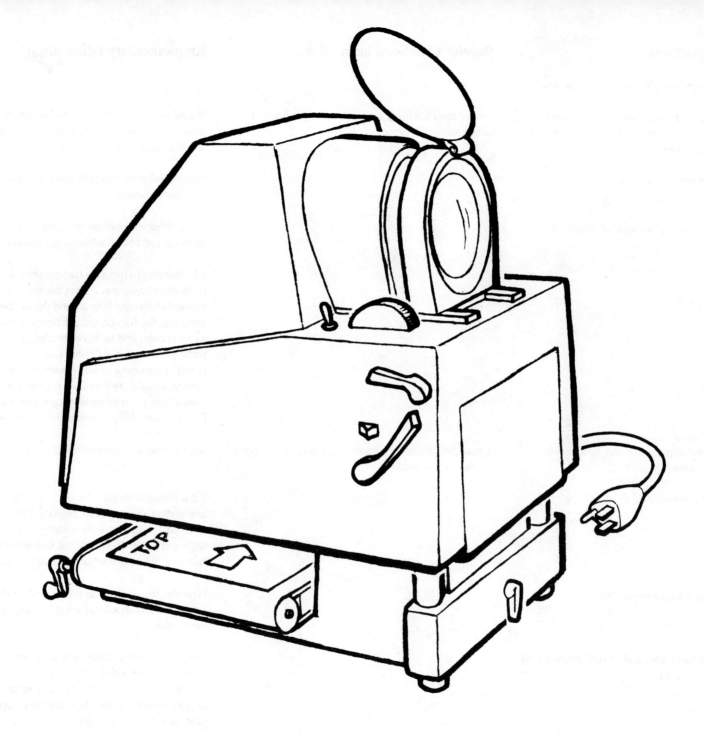

Opaque Projector

The opaque projector reflects an image of materials onto a screen or other surface. This machine is capable of projecting enlarged images of graphic materials such as books or photographs, and small 3-dimensional objects in natural color. Flat materials up to 10 inches square may be projected. Larger flat materials may be projected in sections.

The chief disadvantage of this projector is its inefficient light output. A very dark room is necessary for utilization. High temperature of the bulb will damage heat sensitive materials.

Images may be projected onto large sheets of paper with the opaque projector to create large display materials by selective tracing.

Objectives

Observable Behavior

1. Set up projector and project opaque material.

2. Use pointer effectively.

3. Restore machine to storage conformation.

Conditions

Given an assigned page in a textbook. Time limit 1 minute.

Given an assigned word on the textbook page. Time limit 15 seconds.

Time limit 45 seconds.

Level of Acceptable Performance

The image must fill the screen area, be level horizontally, and in focus.

Pointer arrow must be accurately positioned on material.

Lens covered, legs retracted, platen area closed, cord wound neatly.

Essential Parts

Typical

Access door
Heat-absorbing glass
Platen
Power cord
Controls:
    Elevation
    Focus
    Platen
    Power

Additional on Beseler

Lens cover
Controls:
    Platen lock
    Pointer

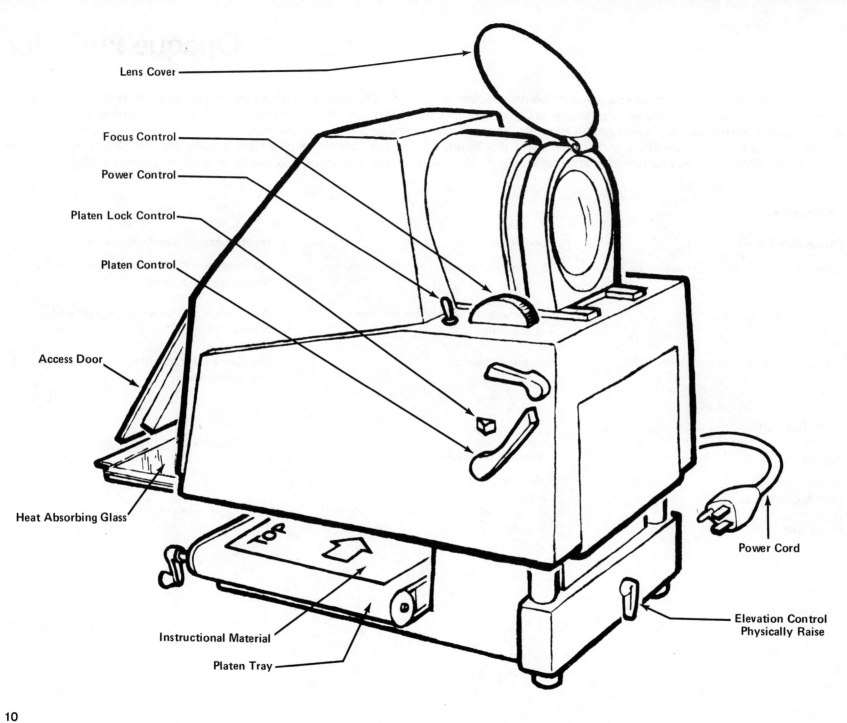

General Instructions

1. Read Projection Principles before proceeding.

2. Tie *power cord* around the bottom of projection stand leg and connect to suitable power source.

3. Raise *lens cover* if provided.

4. Open platen area and insert *instructional material* to be projected.

5. Close *platen.*

6. Turn lamp **on.**

7. Adjust position of projector, *focus control,* and *elevation control* until a satisfactory image is obtained. Physically raise on legs.

Beseler Opaque Projector Instructions

Hold *platen lock control* forward while moving *platen control* in clockwise direction.

Lowering projector will bring *heat-absorbing glass* (if it has not been broken or removed) down onto *instructional material.*

The *power control* has two active positions: **fan only** and **fan and lamp.**

Elevation control unlocks legs to facilitate raising the front of projector.

Supplementary Information

The *lens cover* should always be positioned over lens to protect it from dirt and damage when projector is not being used.

On some projectors, *platen area* is opened with a lever at rear of machine. The top of copy should be toward the rear of the projector. When placing a book in machine, it may be necessary to place a block or another small book under the side with the least pages in order to facilitate focus. The *heat-absorbing glass* will hold the pages of a book in the same focal plane and keep them from fluttering due to cooling fan.

Removal of continuous-belt *platen tray* and the *heat-absorbing glass* will enable user to insert larger 3-dimensional objects through *access door*.

On some machines there is one active position that combines fan and lamp.

Physical effort must be exerted to raise projector. The *elevation control* locks it in desired position.

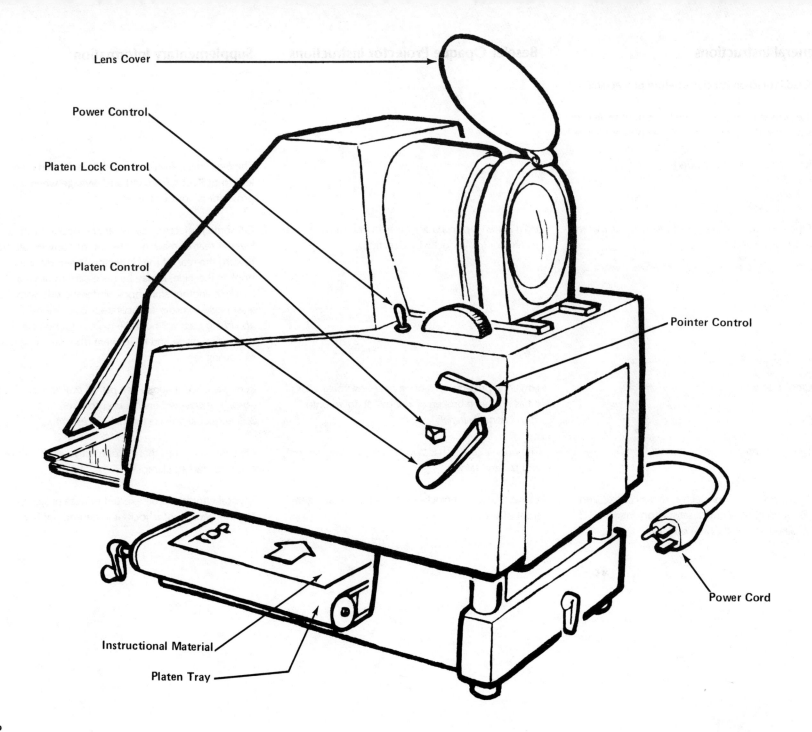

General Instructions

8. Use *pointer control* if appropriate.

9. Turn projector lamp **off.**

10. Remove *instructional material* from projector.

11. Turn off cooling fan, disconnect *power cord,* and return projector to storage conformation.

Beseler Opaque Projector Instructions

The pointer is activated by manipulating the control up or down, in or out. Exact location of arrow is determined by watching screen.

Move *power control* to **fan.**

Move *platen lock control* forward to permit raising projector with *platen control.*

Move *power control* to **off.** Lower projector onto *platen tray,* lower *lens cover,* and withdraw legs to storage position.

Supplementary Infomation

On some projectors, slide the lever to one side to lower *platen tray.*

Multiple-load Slide Projector

Multiple-load 2 × 2 slide projectors accept transparencies produced on 35mm, 126, and 110 slide film. Circular and rectangular storage trays provide the user with a means of packaging sequences of slides for use by individuals or large groups. Presentation of the slides may be manipulated remotely by the user or by electronic control devices.

Some types of multiple-load 2 × 2 slide projectors have the capability of projecting filmstrips with an optional attachment. A popular carousel-type projector, such as the Ektagraphic, accepts trays with a capacity of up to 140 slides. An optional stack-loader provides the capability of conveniently viewing a stack of up to 40 slides. With appropriate extension cords, the user may be 35 feet away from the projector and trigger slides as desired with a remote control device. An automatic sequence timer makes it possible for the projector to be set up to present slides at regular intervals until it is stopped.

Objectives

Observable Behavior

1. Set up projector to show a group of slides.

2. Slides returned to box. Projector restored to storage conformation.

Conditions

Given a set of slides; some with thumbspots and others without. Time limit 2 minutes.

Time limit 2 minutes.

Level of Acceptable Performance

Slide tray positioned to move slides through machine, focused and positioned to fill the screen width.

All slides to be removed from carrier; projector leveled and power cord stowed.

Essential Parts

Typical

Lens
Power cord
Slide tray
Controls:
    Elevation
    Focus
    Power
    Slide change

Additional on Kodak Ektagraphic III AT

Cord compartment
Gate index
Lock ring
Plug orientation mark
Remote control jack
Slide compartment
Slide identification number
Slide slot

Controls:
    Auto timer
    Leveling
    Remote
    Select
    Forward
    Reverse
    Auto focus
    Quick release

Available on Some Models

Tray release

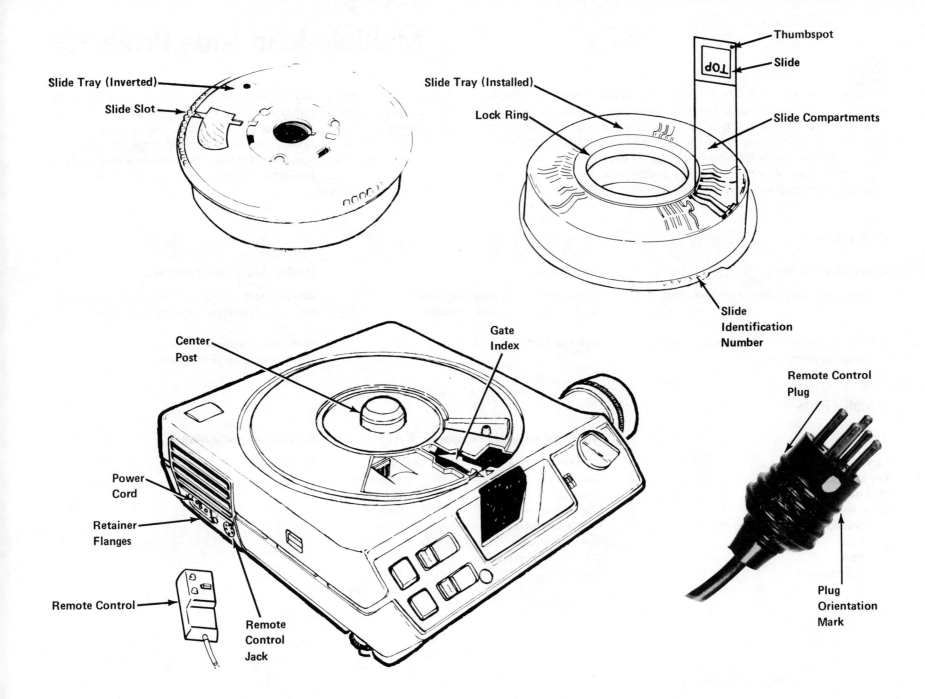

| **General Instructions** | **Kodak Ektagraphic III AT Instructions** | **Supplementary Information** |

Projecting 2 × 2 Slides

1. Read Projection Principles before proceeding.

2. Tie *power cord* around bottom of the projection stand leg and plug into a suitable power source.

   The 3-wire *power cord* is permanently attached and is stored on *flanges* on bottom of projector. Unwrap as much as needed and leave the rest on *retainer flanges*.

   A permanently attached 3-wire power cord is stored in a compartment on the bottom of some projectors. The cord should be pulled out to its full length before turning projector on, to allow proper air circulation through projector. Some projectors have separate power cords.

3. Plug *remote control* into *remote control jack*.

   The white *plug orientation mark* should be **up.** The color of orientation mark denotes number of wires in cord. *Remote control* units should not be interchanged among other models of carousel-type projectors.

   A white mark indicates a 5-wire cord, a red mark 4 wires, and a yellow mark 3 wires, on Kodak projectors.

4. Load *slide tray*.

   Move *slide slot* on bottom of *slide tray* to *slide identification number* **0.** Turn tray right side up with **0** in the 3 o'clock position and disengage the *lock ring* by moving it counterclockwise while holding tray stationary. Arrange slides consecutively in numbered *slide compartments*. When all slides are in tray, replace *lock ring* by turning it clockwise until a detent action is felt or it cannot be rotated further. Secure the *lock ring* to prevent slides from falling out if tray is inverted. Numbering slides is an additional preventive measure for retaining correct order.

   For multiple-load projectors, do not use bent slides or any that do not move easily into *slide compartments*. Handle all slides by edges only. To load slides hold in normal viewing position and rotate 180 degrees. Slides should enter projector in upside-down position. Loading slides is facilitated by adding *thumbspots* before use. *Thumbspots* should be placed in lower left-hand position (typically the glossy, nonemulsion side will be toward viewer). Slides should be loaded into tray with *thumbspot* in upper right-hand corner toward next larger number on tray.

5. Install *slide tray*.

   Hold *slide tray* over projector, align it with *center post* on top of projector, and rotate to place *slide identification number* **0** at *gate index*. Lower slide tray and seat it firmly. If *slide tray* and projector do not seem to mesh, recheck to determine that *slide slot* is at **0.**

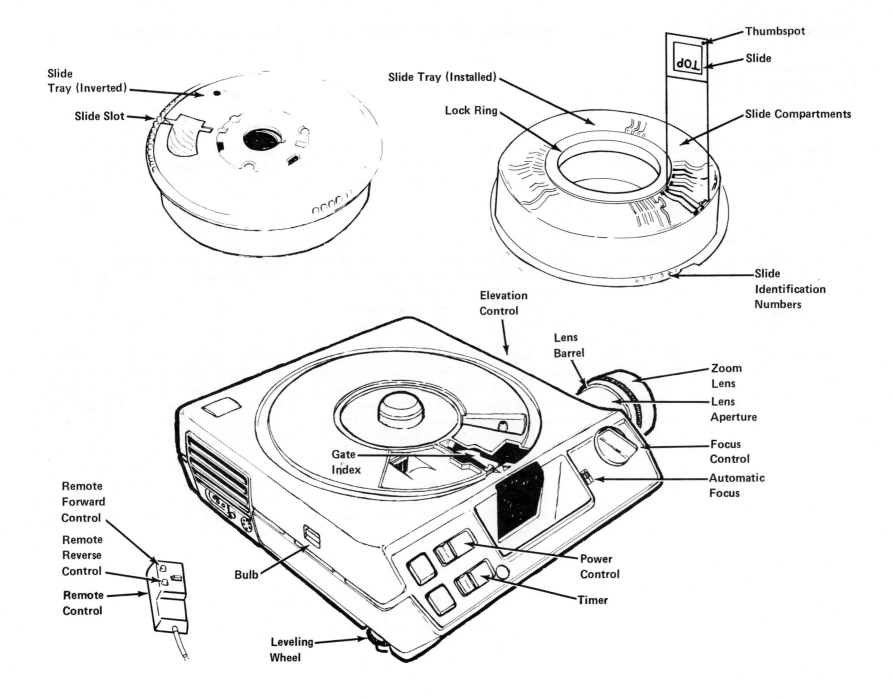

General Instructions	Kodak Ektagraphic III AT Instructions	Supplementary Information
6. Insert *lens* if it is not already in position.	Projectors in semipermanent locations may have a *lens* that is not ordinarily removed for storage. Push *lens* into projector past spring pressure with toothed side of *lens barrel* toward lower right side of *lens aperture*. The teeth in the side of *lens barrel* must engage gear teeth of *focus control* to move *lens* for critical focusing. Rotation of *focus knob* will move *lens* into *lens aperture*.	On some projectors, the *focus control* must be moved toward right side of projector to release focusing mechanism and permit insertion or removal of *lens*. A *zoom lens* is furnished with some projectors to permit adjustment to varied conditions.
7. Set *automatic timer control* at **off.**	The *automatic timer control* may be adjusted for **off** (manually) or **"F . . . S"** (automatic) projection intervals.	Some *timers* may be set at a given number of seconds that each slide will be projected.
8. Turn projector **on**.	Move *power control* to either **lo** or **hi.**	Use **lo** unless **hi** is necessary for adequate brightness because the *projection bulb* will last much longer when **lo** setting is used.
9. Advance first slide and adjust *elevation control*, image size, and *focus control*.	Push *forward control* or *remote forward control*. Depress quick release *elevation control* to provide rapid height alignment. The *leveling wheel* provides precise adjustment.	*Slide identification numbers* on *slide tray* indicate which slide is aligned with *gate index*. If projector is equipped with a *zoom lens*, adjust lens by rotating *lens barrel* so that image fills screen area; then focus. If projector is equipped with *automatic focus option*, it is necessary to focus only the first image with *auto focus switch* in **off** position; remaining images will be adjusted automatically. On some models, forward and backward movement of remote *focus control* will override *automatic focusing mechanism* until control is released, then automatic focus is reinstated.

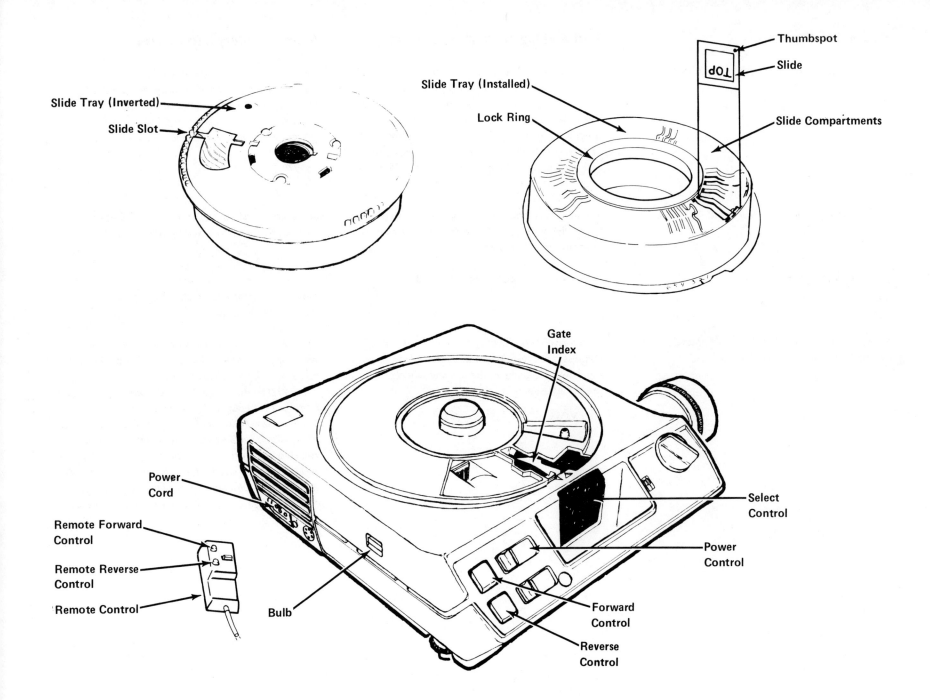

| **General Instructions** | **Kodak Ektagraphic III AT Instruction** | **Supplementary Information** |

10. Advance to subsequent slides and focus as necessary. It may be appropriate to skip some that have been previously shown.

Depress *remote forward control*. Constant pressure on button will cause intermittent advance of *slide tray* to succeeding numbers. *Remote reverse control* may be used to project next lower slide identification number. On some models, slides may be advanced or backed up by depressing *forward* or *reverse controls*. The *select control* may be used to release *slide tray* and allow selective projection of slides by rotating *slide tray* to desired position. Hold *select control* while rotating *slide tray*. Release of *select control* will immediately project slide in slot opposite *gate index*.

On some models of multiple-load slide projectors, projection sequence cannot be reversed, nor can selected slides be conveniently skipped.

Replacing Bulb

11. *Bulb* replacement procedure should be known.

Bulb may be replaced from rear of projector without disturbing alignment with screen.

On some models of multiple-load slide projectors, the projector must be turned upside down to replace *bulb*.

Stowing

12. Remove *slide tray*.

Depress *select control* and rotate *slide tray* to **0** at *gate index*. Lift *slide tray* from projector. Release *select control*.

Operation of *select control* to remove *slide tray* on some projectors may require *power control* to be in an active position.

13. Move *power control* to **off.**

14. Remove slides from *slide tray*.

Turn *slide tray* right side up and disengage *lock ring* by moving it counterclockwise while holding *slide tray* stationary. Replace *lock ring* after removing slides.

15. Disconnect *power cord* and *remote control*. Return them to storage.

Replace projector *cover* if provided.

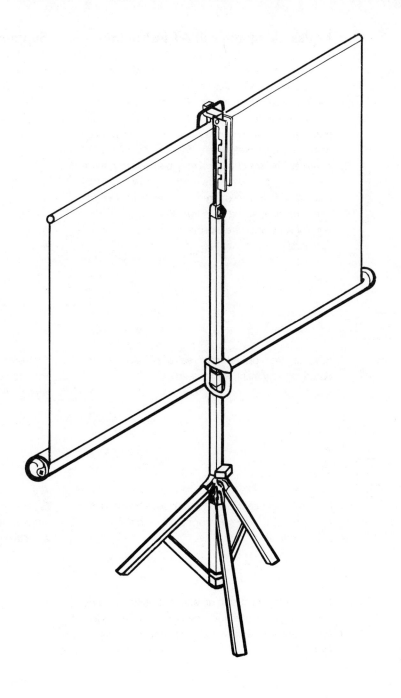

Portable Screen

The portable tripod screen described in this book is typical. Portable screens may be positioned as needed and offer a degree of flexibility that is not possible with wall-mounted screens. The choice of screen surface is discussed in Projection Principles. These screens are relatively simple to set up, but the inexperienced user frequently is not conscious of the effects of gravity and spring tension.

The photographics are the property of Da-Lite Screen Co., Inc., Warsaw, Indiana.

Objectives

Observable Behavior

1. Set up the portable tripod screen for use.

2. Restore screen to storage conformation.

Conditions

Given a screen that has been collapsed for transport or storage. Time limit 3 minutes.

Time limit 2 minutes.

Level of Acceptable Behavior

Screen extended to maximum height and legs fully extended.

Screen housing fastened in upright storage position and legs retracted.

Essential Parts

Typical

Center post
Screen hanger
Screen housing
Transport handle
Tripod legs

Additional on Da-Lite

Center post extension
Center post hook
Storage aperture
Storage pin
Controls:
    Center post release
    Leg release
    Screen housing position

Available on Some Models

Keystone eliminator

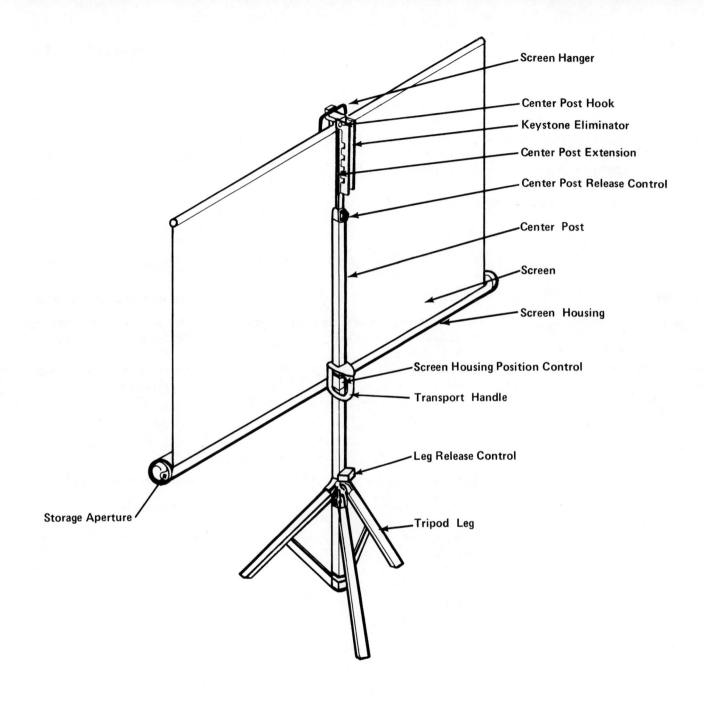

| **General Instructions** | **Da-Lite Screen Instructions** | **Supplementary Information** |

Setting Up Screen

1. Extend *tripod legs*.

   Grip *transport handle* and suspend unit with one hand while pushing *leg release control*.

   Some tripod screens have *leg release controls* at the point on *center post* where *tripod legs* are collapsed; others have a control at bottom end of *center post*.

2. Place unit, with *tripod legs* fully extended, on floor.

   Position unit away from open windows where it may catch a breeze and tip over. Also avoid positions in moving traffic because in subdued light conditions, someone might trip over *tripod legs*.

3. Disengage *screen housing* from *center post* and swing it to horizontal position.

   Tripod legs are positioned so *screen housing* will swing only one way after it is unlocked by pulling *center post release control* and raising *center post extension* far enough to clear *storage aperture*.

4. Pull *screen hanger* high enough to attach it to *center post hook*.

5. Extend *center post extension* and *screen*.

   Pull *center post release control* with one hand and raise *center post extension* and *screen* with other.

6. Adjust *screen housing position control* to move bottom of *screen housing* higher on *center post* and then repeat step 5.

   Screen may be adjusted to reflect an image projected over heads of audience if projector is also elevated.

7. Adjust to eliminate keystone effect if possible. Horizontal accommodation is made by adjusting the position of legs.

   If a *keystone eliminator* is provided, extend it forward at the top of the screen and move the *screen hanger* appropriately forward.

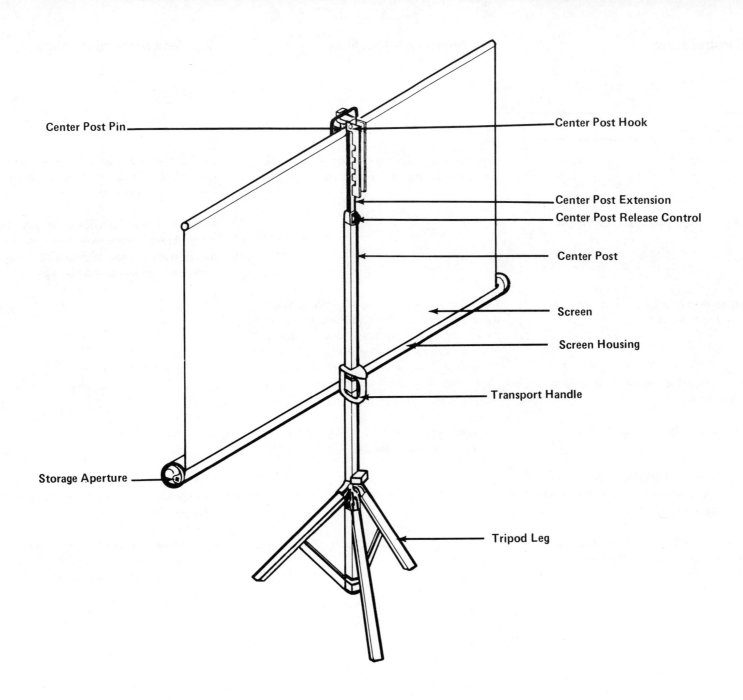

General Instructions	Da-Lite Screen Instructions	Supplementary Information

Stowing

8. Lower top of *screen*.

   Grip *center post extension* firmly between *center post release control* and top of *center post extension* with one hand and pull *center post release control* to lower top of *screen*. Roller tension on *screen* will cause it to drop rapidly, so grip *center post extension* firmly.

9. Roll *screen* into *screen housing*.

   Lift *screen hanger* from *center post hook*.

   Retract *screen* slowly to prevent damage.

10. Swing *screen housing* to vertical position and secure it.

    Screen housing will swing only one way. Pull *center post release control* and lock *center post pin* to *storage aperture* to allow convenient transport.

11. Collapse *tripod legs*.

    Suspend *screen* by *transport handle* and grasp the top flange of *tripod legs*. Pull along *center post* until *tripod legs* are secured at base of *center post*.

Chapter II Audio Principles

A basic principle of audio reproduction is that sound is created by a movement of air molecules. Sound moves from a vibrating source in all directions. It is affected by the:

A. peculiarities of the original sound
B. sound absorbing or reflecting qualities (acoustics)
C. capacity of the recording or playback system to reproduce the original vibration frequency accurately (called fidelity)
D. operator's experience in using the equipment

Most sound reproduction equipment incorporates the following components:

A. a source, which may be live or prerecorded
B. a transducer to change energy from one form to another
    (1) microphone (mechanical vibrations to mechanical impulses)
    (2) speaker (electrical impulses to mechanical vibrations)
C. an amplifier to strengthen electrical impulses
D. a transport system

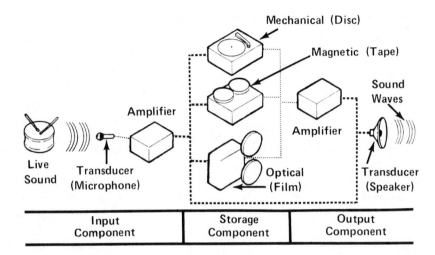

BASIC AUDIO SYSTEM

The open-reel tape recorder allows the greatest ease and flexibility for editing purposes. However, today it is difficult to find open-reel tape recorders for less than $750 because consumer demand is focused on audiocassette and audio microcassette equipment. Cassette recorders may be purchased for half-track monaural or quarter-track stereo applications. Microcassette recorders are pocket-size. The microcassettes used in them are also much smaller than the usual audiocassette.

Intended use for any tape recorder should determine whether it is:

A. monaural (one channel and one input)
    (1) full track (professional equipment)
    (2) half track
    (3) quarter track
B. stereo (two or more channels and two or more inputs)
    (1) half track
    (2) quarter track

The number of tracks on a tape is determined by the recorder used, not by the tape selected. Audiotape is produced on several kinds of backing material in a variety of thicknesses, which affects the length and quality of recording. Very thin backing material capable of long recording or playback potential would be .5 mil in thickness. At the other extreme, a heavy backing material would be 1.5 mil in thickness. The thinner the backing material, the more tape can be wound on a reel. The total playing time for a cassette is indicated by the number on the label, e.g., a C60 will play 60 minutes or 30 minutes on each side. Microcassettes are labeled RT-60 MC (recording time of 60 minutes) and will accept 30 minutes on each side. Most microcassette recorders have two recording speeds, SP for standard speed and LP for long play. LP speed provides a total of 2 hours of recording time on an RT-60 MC.

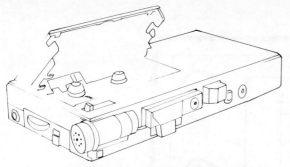

MICROCASSETTE RECORDER

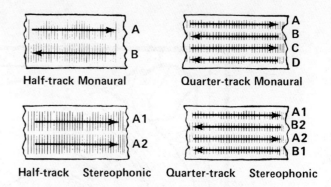

AUDIOTAPE CONFIGURATIONS

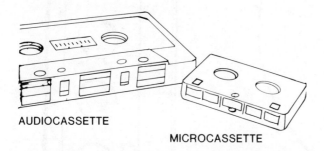

AUDIOCASSETTE   MICROCASSETTE

The speed at which a tape recording is made affects the fidelity of the sound and the total time for recording. High speed ordinarily produces high fidelity as well as high consumption of tape.

Audiotape can be edited by physically cutting the tape at a 45 degree angle and splicing without overlapping to facilitate passage through the machine. A special splicing tape is used that will not deteriorate or gum up the recorder with adhesive residue. This editing process is easier when using open-reel tape found in professional applications. It is possible to splice cassette tape, but few people have the patience and digital dexterity to do it properly. It is much easier to edit by connecting two recorders with a patch cord to transfer selected signals from one or more existing tapes onto a new tape. This process is often called *dubbing*. Both processes are relatively simple, and professional results are easily achieved. An accessory patch cord may be used to record signals from a radio or other source by inserting one end into the auxiliary input jack on the recorder and connecting the other end to the speaker wires or output jack on the source. In this way, room noises that would be picked up by a microphone are eliminated.

Always connect inputs to outputs, or outputs to inputs. Observe labeled jacks on the equipment and think about the desired directions of the impulses.

Most tape recorders are equipped with a detachable microphone to allow remote placement, which reduces the likelihood of recording motor noise. A remote microphone may be cushioned with soft materials to improve fidelity. A test recording will help to determine appropriate microphone placement and volume adjustment. Microphones are designed to serve a variety of purposes and are designated:

A. unidirectional — picks up sound primarily from one direction, used to eliminate ambient (unwanted) sound
B. bidirectional — picks up sound from opposite directions, used for recording face-to-face interviews
C. omnidirectional — picks up sound from all directions, used for group interaction

Record Player

Many types and brands of record players are used in schools. There is a wide selection of disc recordings available to the classroom teacher which may be used effectively for a variety of instructional purposes. Standard (78 rpm), microgroove (45, 33 1/3, 16 2/3 rpm) monaural, and stereo records can be played on record players designed to accommodate all types. Stereo records may be safely played only on monaural machines equipped with a safeguard needle.

Objectives

Observable Behavior

1. Set up record player and play the disc recording provided.

2. Restore record player to storage conformation.

Conditions

Given a suitable disc recording. Time limit 2 minutes.

Time limit 2 minutes.

Level of Acceptable Performance

Turntable speed, tempo control, needle, volume, and tone controls properly adjusted.

Pickup arm secured, speaker and power cords stowed, and speaker cover closed.

Essential Parts

Typical

Needle selector
Needle (stylus)
Pickup arm
Pickup arm retainer
Power cord
Speaker
Turntable
Controls:
    Power
    Speed
    Volume/phono

Additional on Newcomb

Power cord compartment
Tempo indicator
Controls:
    Bass tone
    Tempo
    Treble tone

Available on Some Models

Automatic record changer
Controls:
    Pickup arm balance

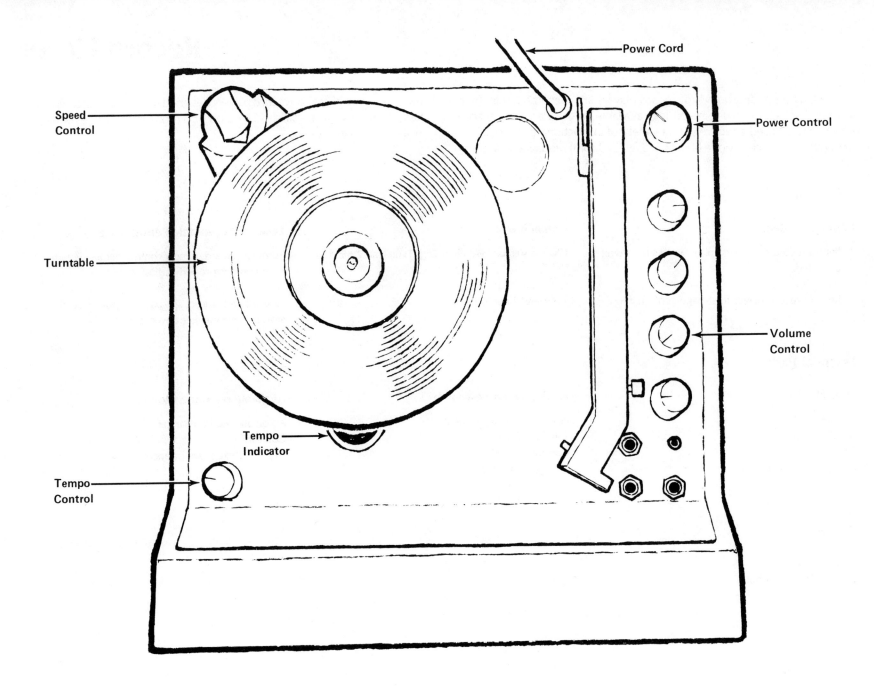

General Instructions   **Newcomb Record Player Instructions**   **Supplementary Information**

Playing a Record

1. Read Audio Principles before proceeding.

2. Remove or open cover.

3. Tie *power cord* around bottom of cart leg and connect to suitable power source.

4. Turn *power control* **on.**

                                                                                             The *amplifier* or *motor control* may be independent from or combined with *volume control*.

5. Connect *speaker plug* to *speaker jack*.

                                                                                             On some makes or models, the *speaker* is built into the player and does not require external connection.

6. Set *speed control* to speed indicated on record label.

                                                                                              The three common speeds are 33 1/3, 45, and 78 rpm (revolutions per minute).

7. Adjust *tempo control* if there is one on machine and then switch *motor control* **off.**

                                        Observe 4 rows of dots through *tempo indicator*. If *speed control* is set for 33 1/3 rpm, the second row of dots (from the outside) should appear to stand still. Adjust *tempo control* until this row of dots appears to remain stationary.   This adjustment insures precise *turntable* rotation speed to compensate for electrical or mechanical variables.

8. Place record on *turntable*. Switch *power control* **on.**

                                                                                              Handle records by edges to prevent damage. If there is an adjustment for size of record, such as 7, 12, or 16 inch records, determine that it is properly set.

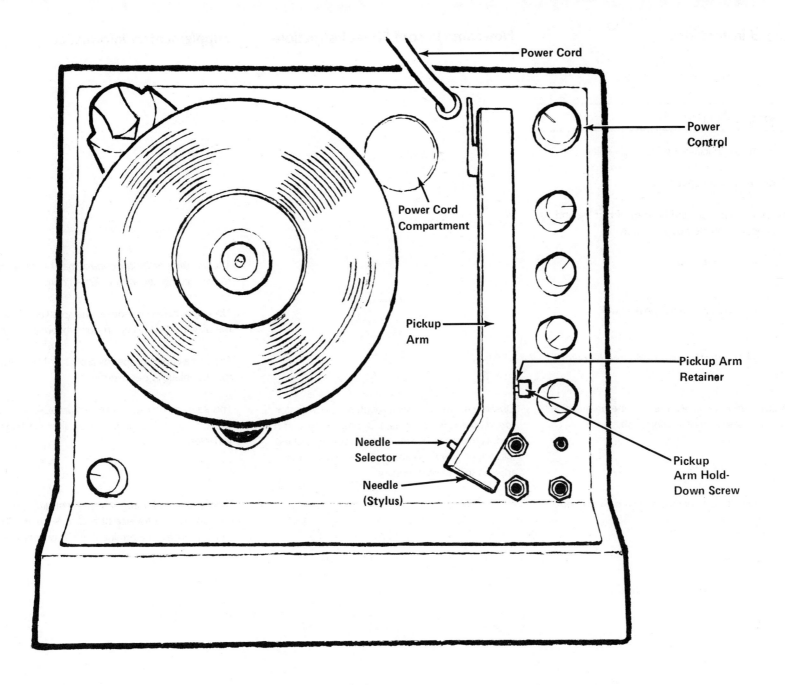

| **General Instructions** | **Newcomb Record Player Instructions** | **Supplementary Information** |

9. Set *needle* (stylus).

Rotate *needle selector* at end of *pickup arm* to desired position.

Adjust pickup arm balance weight to the least pressure that will cause needle to stay in grooves of record.

10. Release *pickup arm* from *pickup arm retainer* and position *needle* in outer groove of record.

Loosen *pickup arm hold-down screw.*

11. Adjust *volume/phono* and *tone controls* to suit environment.

Move to farthest points of room to determine proper adjustment.

12. Remove *pickup arm* from record and secure it to *pickup arm retainer.*

Set *pickup arm* in clamp and tighten *pickup arm hold-down screw.*

The *pickup arm* must be secured any time machine is likely to be moved. This precaution will prevent damage to *needle.*

13. Switch *power control* to **off.**

On some machines, the two are combined in one control.

Stowing

14. Remove record from *turntable* and return it to protective envelope immediately.

15. Disconnect *power cord* and return machine to storage conformation.

Stow *power cord* in *power cord compartment.*

If *power cord* plug is brought to *power cord compartment,* and length of cord is halved several times, a small enough bundle may be created to be inserted into *power cord compartment.*

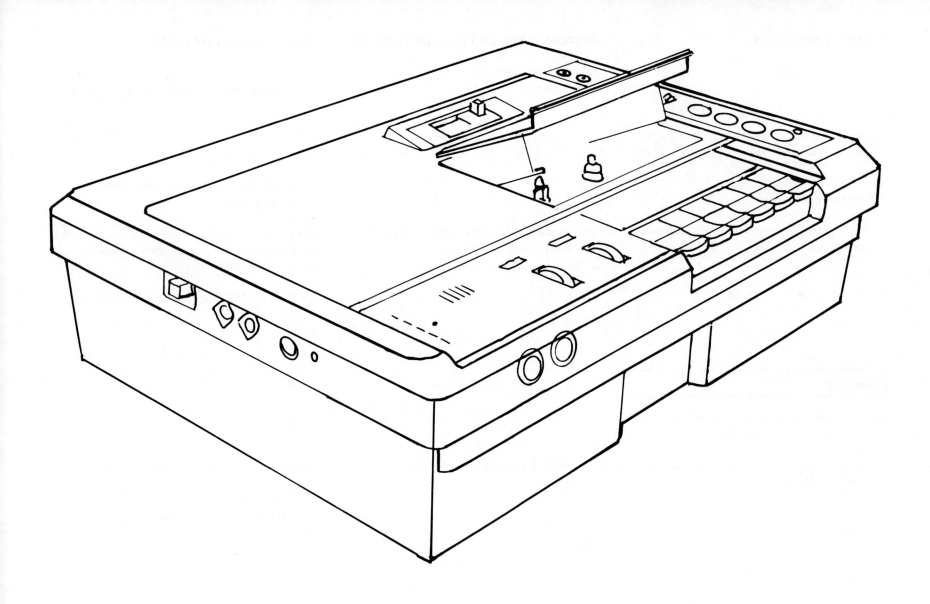

Audiocassette Recorder with Slide/Sync System

Portable audiocassette tape recorders provide compact convenience for amateur and professional applications. Anyone can produce audio materials for classrooms or personal use. Most cassette recorders have simplified controls, and some have automatic level controls to insure adequate sound levels under normal recording conditions. Each recording session should be evaluated carefully to determine whether manual or automatic control will produce desired results. Recording units may be purchased for as little as $50.

The additional advantages of programming for two-projector presentations and slide/sync capability are found on more expensive models. For classroom use, a 4 3/4 inch speaker and minimum 5 watt output power is recommended. Smaller units are available for individual use. Microcassette audio recorders are a convenient vehicle to record memos, interviews, and on-site sounds. They range from $50 to several hundred dollars in price.

Objectives

Observable Behavior	Conditions	Level of Acceptable Performance
1. Set up recorder and make a recording.	Given a cassette tape. Time limit 2 minutes.	A clear, 30-second recording of operator's voice.
2. Rewind and play back tape recording.	Time limit 1 minute.	Playback with properly adjusted volume and tone controls.
3. Restore recorder to storage conformation.		Function controls should be left in stop position. Additional mike and power cord disconnected.

Essential Parts

Typical

Access door
Auxiliary input jack
Battery and AC cord storage
Battery power indicator
Built-in condenser mike
Cassette compartment
External mike jack
External speaker/headphone jack
Record level indicator
Safeguard interlock
Safety tabs
Spindles
Tape position indicator

Additional on Sharp RD-671AV

Controls:
    AC/DC sel
    Auto level
    Fast forward/cue
    Pause
    Play
    Record
    Rewind/review
    Stop
    Counter reset
    Tape position reset
    Tone

Controls:
    One KHz
    One-hundred fifty Hz
    Program restart
    Public address
    Sync mode
    Sync pulse record
Program restart remote jack
Projector#1 output jack
Projector#2 output jack
Remote on/off jack
Sync input jack
Sync output jack
Sync pulse record indicator

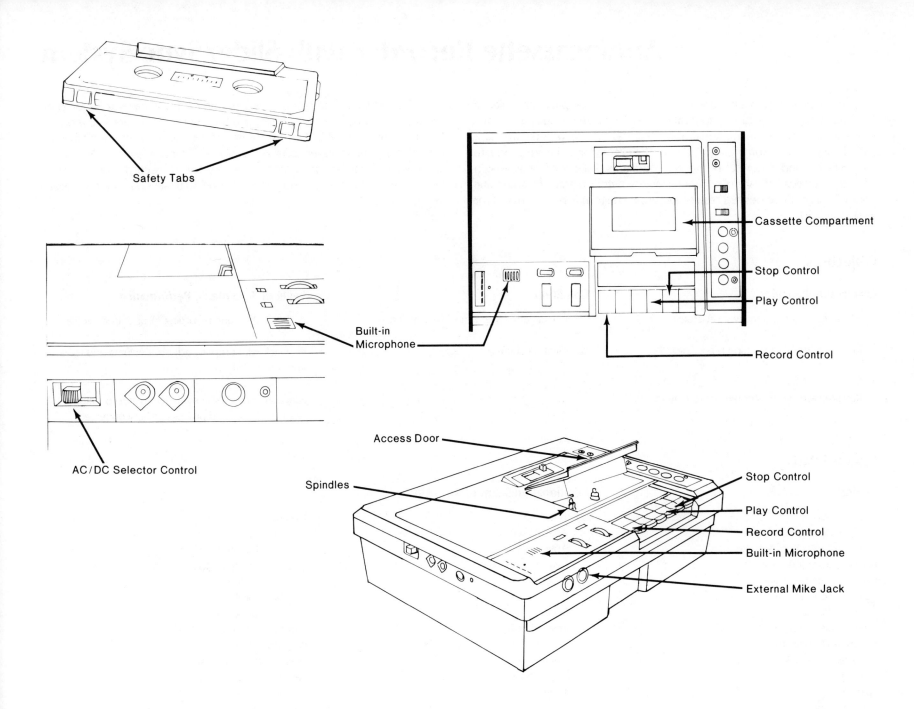

| **General Instructions** | **Sharp RD-671AV Audiocassette Tape Recorder Instructions** | **Supplementary Information** |

Making a Recording

1. Read Audio Principles before proceeding.

2. Check batteries for DC operation, connect to suitable power source. Select appropriate power source at *AC/DC selector control*.

    Insert AC power cord into external power source. To test for battery charge level, open *access door* by lifting front edge, depress *play control,* and check for *spindle* rotation. Only the take-up *spindle* will rotate. Lack of rotation indicates need for battery replacement.

    A frequent visual check should be made to verify that batteries are not leaking. Batteries that are discharged should be removed immediately to prevent damage to recorder. Weak batteries will cause slowdown of transport system resulting in distorted playback.

    NOTE: Batteries should be removed when using AC power source due to possibility of ignition of gases from weak, leaking batteries.

3. Stop recorder.

    Depress *stop control.*

4. Insert *cassette*.

    Hold *cassette* with full reel of tape on left side. Either tape or leader should be exposed through slots in near edge. Far edge should be inserted first. Door to cassette compartment should close easily.

    Procedures for cassette insertion vary; however, it will seat properly only in correct position.

5. Connect *microphone* to recorder.

    The Sharp RD-671AV has a built-in *microphone*. An external microphone may be plugged into the *external microphone jack.*

    If there is a switch on microphone, there will be a second plug on cord that should be inserted into remote jack. As a recording is made, previous sound impulse pattern on tape is automatically erased. It is possible on some machines to accidentally erase a program. To prevent this from happening to valuable recordings, many machines are equipped with a *safeguard interlock*. On the back of the cassette are two small *safety tabs* which, when removed, allow the *safeguard interlock* to move into a position that prevents the *record control* from being activated. In operating position, the absence of the tab at left rear will protect the material from being erased and rerecorded.

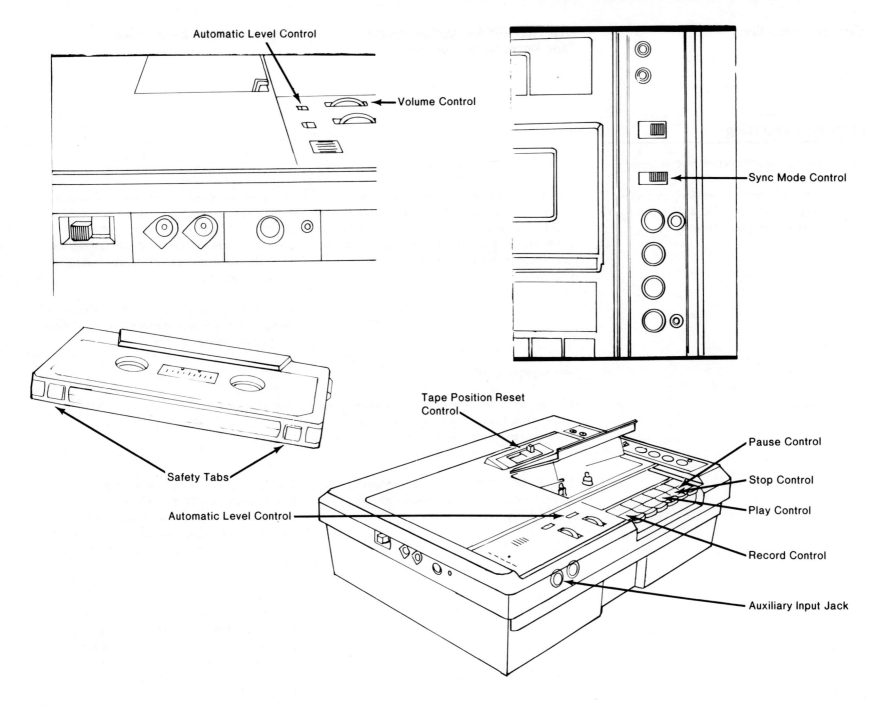

General Instructions	**Sharp RD-671AV Audiocassette Tape Recorder Instructions**	**Supplementary Information**
6. Adjust *volume control*.	Two systems are provided, *manual control* and *automatic level control (ALC)*.	*Manual control* may be used when operator can continuously monitor audio level. At proper audio level, indicator (LED or VU meter needle) should average at 0 db mark. *Automatic level control* may be used when operator does not wish to monitor sound level. An objectionable hiss may be heard when amplifier strives to pick up low-level audio.
7. Set *tape position reset control* to **000**.	Depress *tape position reset control*.	Numbers indicate the number of revolutions of take-up reel.
8. Adjust *recorder controls*.	Set *sync mode control* to **off**. Depress *pause control*. Depress *record* and *play controls* simultaneously.	On some machines, *remote on/off control* to start and stop tape motion in both record and play modes is located on mike. When remote control is used, tape will not move unless switch is on. Fast forward and rewind will usually operate regardless of remote switch control.
9. Activate controls.	Release *pause control* by depressing it. Hold *microphone* 6 to 8 inches in front of speaker's mouth or other sound source.	To temporarily stop recorder during taping, depress *pause control*. Depress control again to restart taping. An accessory patch cord may be used to record from a radio or other signal source by inserting one end into *auxiliary input jack* on recorder. This input bypasses recorder's amplifier and produces higher quality sound than using microphone jack.
10. Stop recording when desired or when tape on left reel has been wound onto right reel.	Depress *stop control*.	Many recorders will stop when end of tape is reached. *Stop control* should be used if there is no automatic stop function to avoid unnecessary strain on transport mechanism.
11. Record on the other track by turning cassette over if desired.	Proceed as in steps 4, 7, 8, 9, and 10.	

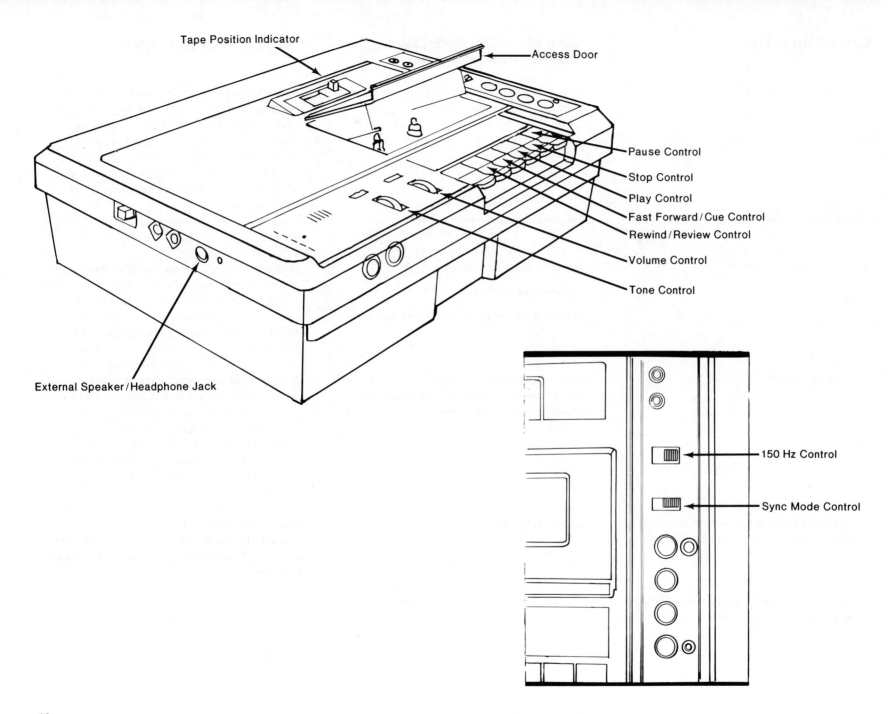

| **General Instructions** | **Sharp RD-671AV Audiocassette Tape Recorder Instructions** | **Supplementary Information** |

12. Rewind *cassette*.

Depress *rewind/review control*. Depress *stop control* when tape is rewound.

13. Stop rewinding when *tape position indicator* reaches **000** if you wish to play back what you have just recorded.

A recorder should never be left in an active mode when tape is not moving. Move all controls to **off**.

Playing a Tape

14. Place *cassette* in recorder.

Lift front edge of *access door*.

15. Play tape. Adjust *volume control* and *tone control*.

Depress *play control*.

In addition to enclosed speaker, a single earphone, an auxiliary speaker, or patch cord to another recorder may also be connected to *external speaker/headphone jack*.

16. Rewind *cassette*.

Depress *rewind control*.

Tape may also be advanced rapidly through a recorder by depressing *fast forward/cue control*. Always return to stop mode if machine does not have automatic controls.

Recording Sync Pulse Track

17. Place prerecorded cassette in recorder.

Lift front edge of *access door*.

Safety tabs on cassette must be in place. A piece of tape over the aperture will prevent safety interlock from shutting off record function.

18. Adjust *sync mode control*.

Set control to separate position.

19. Adjust *150 Hz control*.

Set control to sync.

20. Depress *pause* and *play controls*.

21. Release *pause control*.

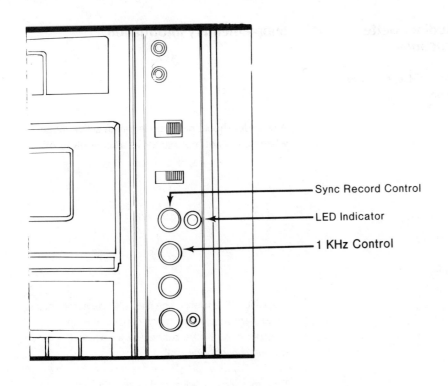
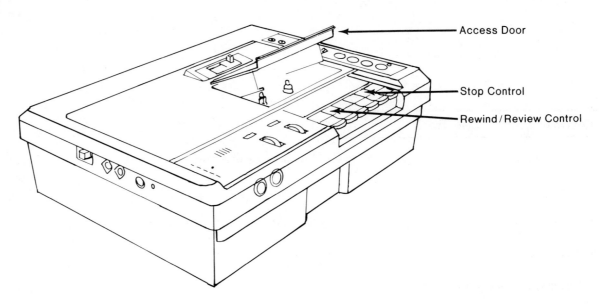

| **General Instructions** | **Sharp RD-671AV Audiocassette Tape Recorder Instructions** | **Supplementary Information** |

22. Press *sync record control*.

The *LED indicator* lamp will light.

23. Advance projector as desired.

Depress *1 KHz control* for each pulse to be recorded on tape.

Sync pulses may be recorded or edited without the recorder being connected to a projector. Each sync pulse may be audible through speaker. It will only be heard during recording, not during playback. It serves as audible verification of sync pulse location.

24. Stop projector.

Depress *stop control*.

25. Rewind tape.

Depress *rewind/review control*.

Stowing

26. Restore recorder to storage conformation.

Lift front edge of *access door* and remove *cassette*.

Chapter III AV Equipment Principles

Popular Visual Formats		Typical Audio Formats
Still		
Slide	*Film Type*	Magnetic Tape
Full Frame	135	Open Reel
		Cassette
Filmstrip		Magnetic Tape
		Cassette
Motion		
Open-Reel Film		Optical Track
16mm		
Videotape		Magnetic Track
3/4″ U-Matic		
1/2″ VHS		
1/2″ Beta		

Coordinated audio and visual stimulus presentations can facilitate the attainment of a variety of educational goals. Generally, it is not appropriate to argue the effectiveness of words versus pictures, or audio versus visual modalities for learning. The combinations should, and usually do, complement each other. Audiovisual presentations can provide vicarious experiences that are considerably more realistic than single-channel presentations. The audio and visual elements reinforce each other through intentional redundancy and thereby accommodate a variety of learning styles. Some learning is most efficiently and effectively served if motion is available to illustrate concepts. With a student population that has gained a major portion of its knowledge from television, educators must recognize that single-channel presentations must be exceptionally fine to maintain student interest and measure up to the competition.

The combinations shown in the table on the left are available commercially, and some are relatively simple to produce locally. Presentation of the still formats is usually accomplished by using projectors and audio equipment discussed in preceding sections of this book. Motion picture projectors and filmstrip projector with audiocassette playback (DuKane Micromatic) are explained in this section because they use both audio and visual channels.

All of the previously discussed General Equipment Operation Principles, Projection Principles, and Audio Principles also apply to the equipment discussed in the chapter and should be reviewed.

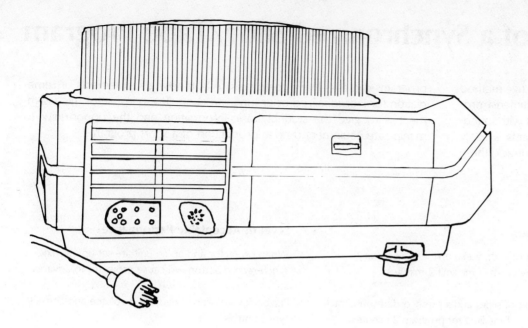

Preparation of a Synchronized Slide/Tape Program

A synchronized slide/tape program is an easy, cost-effective method of media presentation. With limited experience, a basic multimedia production may be created with a slide projector and an audio unit with soundtrack synchronizing capability. Visual material in slide programs is easily modified or updated. By synchronizing the visuals with a soundtrack, these programs can effectively communicate as teaching supplements or remediation for individualized or group instruction. The purpose of this assignment is to give the student basic information and the opportunity to manipulate equipment used in production of a short program.

Objectives

Observable Behavior

1. Set up slide projector and connect with slide synchronizing unit.

2. Synchronize slides with audio.

3. Restore slide projector and audio unit to storage.

Conditions

Given a projector, audio unit, and necessary connecting cords. Time limit 2 minutes.

Given a set of slides and a prerecorded soundtrack on cassette. Time limit for program 2 minutes.

Time limit 2 minutes.

Level of Acceptable Performance

When projector and audio unit are set up, pushing *synch-record* button will cause projector to advance.

Playback with properly adjusted volume and correct sync controls.

All power cords stored properly, cassette tape removed, patch cord stored in cover, and audio unit covered.

Essential Parts

Typical

Audio cassette unit
    AC power
    Cassette platform
    Patch cord
    Patch cord jack to projector
    Sync pulse playback function
    Sync pulse record function

Slide projector
    AC power
    Focus
    Index point
    Lens
    Slide tray
    Sync pulse input jack

Additional on Wollensak

Controls:
    Eject
    Pause
    Power switch
    Record
    Rewind/fast forward
    Stop
    Start
    Stop/restart
    Sync mode
    Tone
    Visual advance

Cord compartment
Index counter

Available on Some Models

Autostop
External speaker jack
Microphone input jack
Record level indicator

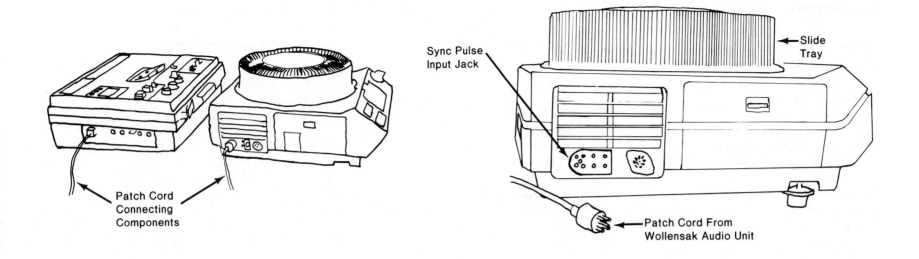

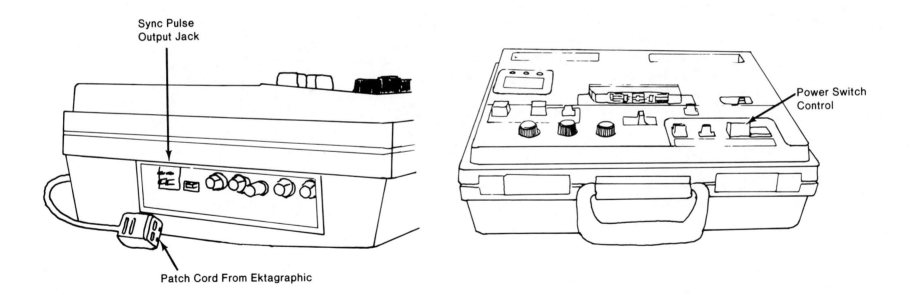

| General Instructions | Wollensak and Kodak Ektagraphic III Instructions | Supplementary Information |

General Instructions

1. Read Projection Principles, Audio Principles, and the section titled Multiple-load Slide Projector.

Setting Up Projector

2. Follow procedure described in the Multiple-load Slide Projector section, steps 2 through 9.

3. After aligning and focusing projector, return slide tray to **0** or the start of the series.

Setting Up Audio Unit

4. Remove cover and connect audiocassette unit to projector.

   Patch cord located in audio cassette unit cover. Connect audiocassette unit to slide projector. Ends of *patch cord* must mate with jacks on each unit.

   Some audiocassette units are capable of synchronizing two or more projectors. Each projector will need a separate *patch cord connection.* If one projector is to be used, it should be connected to *number 1 projector jack*.

5. Tie *power cord* around bottom of cart leg and connect to a suitable power source.

6. Turn power **on.**

   Move *power switch control* to right.

   All recording units do not have a *power switch control.* Recording on this type involves engaging the *recording function,* which automatically turns the unit on.

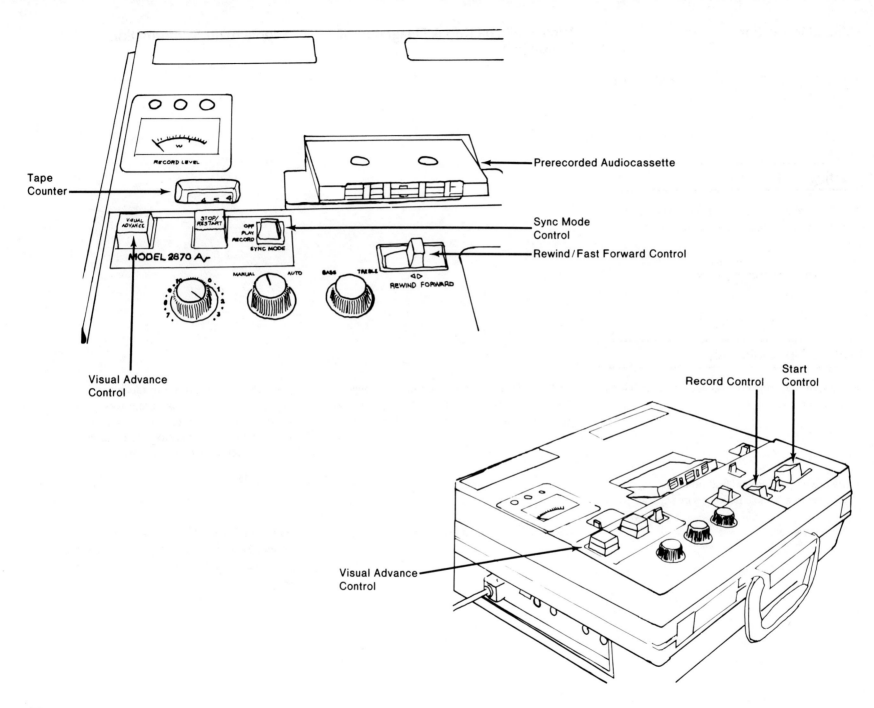

General Instructions

7. Insert soundtrack tape rewound to beginning of program.

Wollensak and Kodak Ektagraphic III Instructions

To record sync pulses on correct area of tape, prerecorded audio cassette must be inserted with soundtrack side up. To fast forward cassette tape, move *rewind/fast forward control* to right. To rewind, move *rewind/fast forward control* to left. All cassette tape movement stops when lever is in center position.

Supplementary Information

Recording units discussed in this chapter use a separate two-track system to record audio and sync signals. One track is used for audio soundtrack, and the sync pulses are recorded on the other. Tape may be played in one direction only because sync pulses cross full width of tape.

Rewind and *fast forward* are also referred to as *review and cue*. To bypass unwanted information on cassette tape, hold down *cue* key until desired area of tape has been reached. Most recording units have a tape counter, which provides easy access to specific areas of a tape if the operator has noted critical areas and remembers to start counter at **000.** The numbers are a record of revolutions of the take-up reel spindle.

Synchronizing Slides and Audio Soundtrack

8. Engage *sync mode record* and *start functions*.

Synch mode control is a three-position switch for recording and playback of synchronized pulses. When *sync mode control* is moved to **record** and the *start control* is depressed, the cassette tape will begin to play prerecorded soundtrack. As tape moves through machine, previously recorded sync pulses are erased. New pulses are recorded by activating *visual advance control*. Button must be depressed to cause it to light. Length of contact does not affect pulse.

Record control is used to record the audio soundtrack portion of program. *Sync mode control* is used to record sync signals necessary to cue projector to advance to next slide. Do not activate *record control* unless the audio track should be changed or erased. *Record control* should be avoided at this point in production unless revision is necessary.

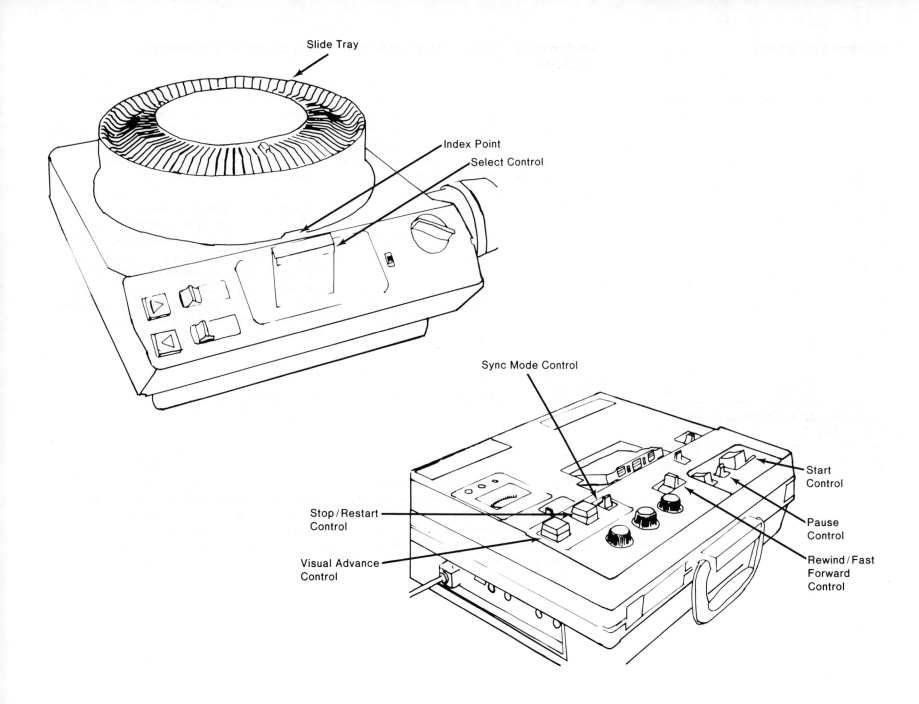

General Instructions

9. *Pause control* may be engaged to halt tape at any point.

10. Record sync pulses on tape.

Wollensak and Kodak Ektagraphic III Instructions

To temporarily stop cassette during either recording or playback, use *pause control*. Pull control toward operator and move to right until it locks in place. To release move control to left.

Depress *visual advance control* to place pulse on tape. *Visual advance control* must light up. If audio unit is connected to projector, the *visual advance control* will light and projector will advance one slide. Depressing *stop/restart control* will place a stop signal on tape and must then be depressed on playback to restart tape program.

Supplementary Information

Pause control is used for making smooth stops and starts during recording as well as for adjusting recording levels prior to starting tape.

Some audio recording units refer to *visual advance control* as *pulse record* or *150 Hz*. In the event of misplaced sync pulses, unit may be stopped, rewound, and played to point prior to error at which time it may be stopped and placed in *sync mode record*.

Playing Back Synchronized Program

11. Depress *stop control*.

12. Rewind cassette to beginning of program.

13. Move slide tray to **0** or beginning of program.

14. Activate *start control* on audio unit.

All audio unit functions will halt. *Sync mode control* will automatically return to *sync mode play*. If *stop/restart control* is lighted, depress control to cancel it.

Move *rewind/fast forward control* to left and hold it until tape is rewound.

Depress *select control* on projector to release tray for rotation. Stop tray when **0** is on *index point*.

Depress *start control* until it locks in down position.

Some machines have keys that may be depressed to rewind tape without exerting constant pressure.

Tape will begin playing soundtrack and signaling projector to advance. During playback, *visual advance control* and *stop/restart control* will light indicating sync pulses that have been recorded on tape.

55

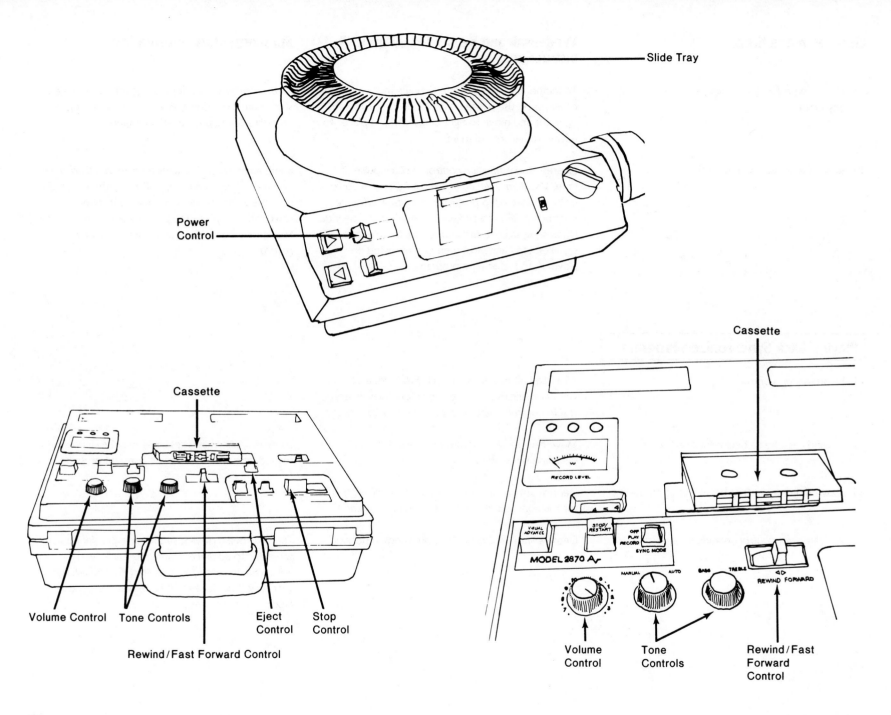

General Instructions	Wollensak and Kodak Ektagraphic III Instructions	Supplementary Information
15. Adjust playback volume on audio unit.	Adjust *tone controls* and *volume control* for listening comfort.	*Tone controls* on most audio recording units operate only during playback. Unless there is a reason for using extreme bass or treble, a midpoint setting is recommended.
16. Stop audio unit when program is completed.	Activate *stop control*.	

Stowing

17. Rewind tape.	Move *rewind/fast forward control* to left.	
18. Remove audiocassette.	Activate *eject control*. Unit must be in stop mode.	
19. Turn audio unit **off.**		Recording units with power control should be left on stop, and *rewind/fast forward control* should be in neutral or center position. Internal parts may be damaged by prolonged storage in an active position.
20. Remove slide tray from projector.	Depress *select control* and rotate tray to **0** before lifting tray off.	Operation of *select control* on some projectors may require *power control* to be **on.**
21. Turn **off** slide projector.		
22. Disconnect patch cord. Return to audio unit cover.		
23. Disconnect *power cords* on audio unit and slide projector. Place in appropriate storage compartments.		
24. Replace audio unit cover.		

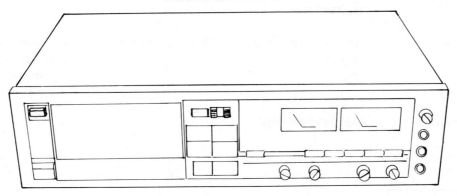
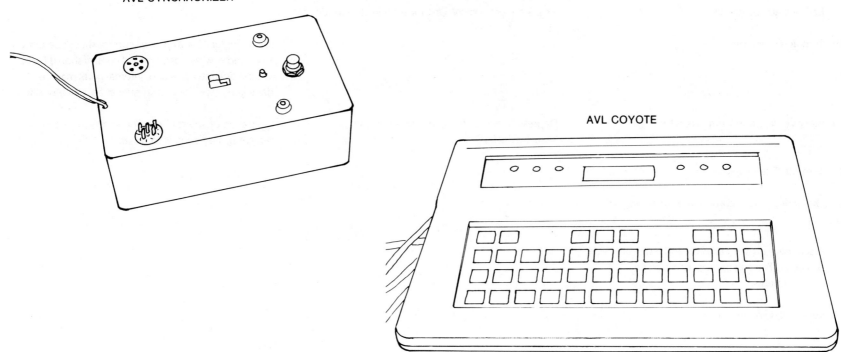

Coyote, Tascam 225 Syncaset, and AVL Synchronizer

The Coyote unit is a memory programmer combined with a 3-projector dissolve unit and is capable of producing simple to complex slide shows. The 225 Syncaset audiocassette tape deck can be used in a variety of ways to produce high-quality audio and also record the control tones for a slide show. This section presents a simple method to use these relatively complex machines to produce a professional-quality show depending on the skill and creativity of the producer. The AVL Synchronizer is used instead of the cue button to put a 1000 Hz tone on the tape each time the sync button is pressed. Refer to the AVL Coyote manual for a more complete understanding of terminology and programming. Refer to the Tascam 225 Syncaset for more information on using it for recording and playback.

Objectives

Observable Behavior

1. Create a 3-projector dissolve show.

2. Present a 3-projector dissolve show.

Conditions

Given a set of slides and a cassette tape with the audio portion of the program. Time limit 8 minutes.

Given a set of slides and a cassette tape with synchronized control tones and audio portion of the program. Time limit 8 minutes.

Level of Acceptable Performance

Program the Coyote with the appropriate commands for advancing slides and synchronize the commands to the audio on one of the cassette tape tracks.

As the audio plays, slides should advance in the projectors as directed by the prerecorded Coyote program commands.

Required Equipment

Coyote dissolve and programmer
Tascam 225 Syncaset tape deck
AVL Synchronizer and patch cord (or Kodak Remote Control)
Two patch cords with RCA connector plugs
Three Kodak Ektagraphic slide projectors
Three slide trays
A 3-stacker tiered slide projector stand
Audio amplifier
AV cart with AC electrical distribution box
Portable screen or white wall
Cassette tape with audio track
Set of slides
Audio script

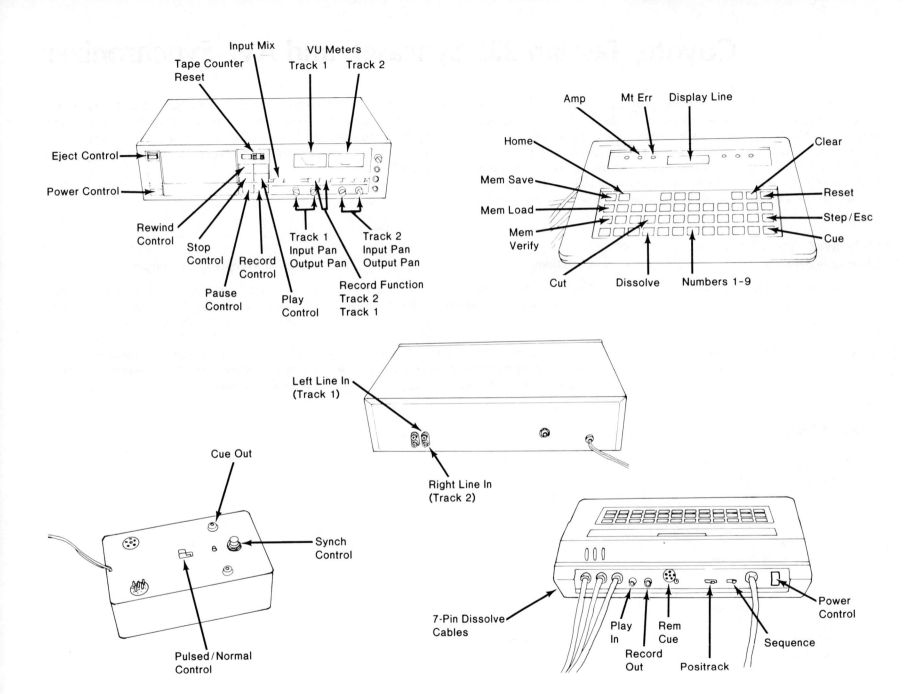

Essential Parts

Syncaset Audiotape Deck

Front panel

Controls:
    Eject
    Input
    Output
    Play
    Pause
    Record function
    Rewind
    Stop
    Tape counter reset

VU meters

Rear panel

    Line in jacks
    Line out jacks

AVL Synchronizer

Controls:
    Pulsed/normal
    Synch
Cue out jack

Coyote

Top front panel
LEDs: AMP OK, MT ERR
    Display line

Keyboard:
    Clear
    Cue
    Cut
    Dissolve
    Home
    Memory load
    Memory save
    Memory verify
    Numbers 1-9
    Reset
    Step/esc

Rear panel

    Controls:
        On/off
        Positrac on/off
        Sequence 2/3
        115/220
Play in jack
Rec out jack
Rem cue jack
Three 7-pin dissolve cables

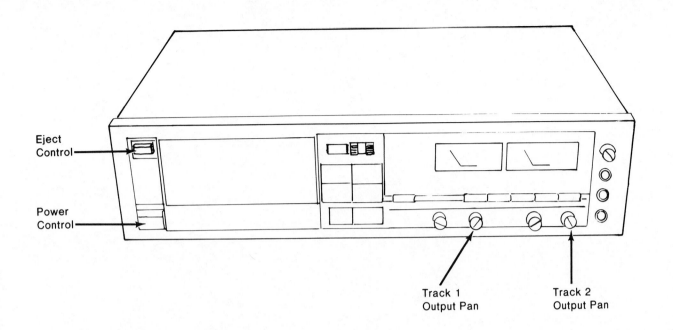

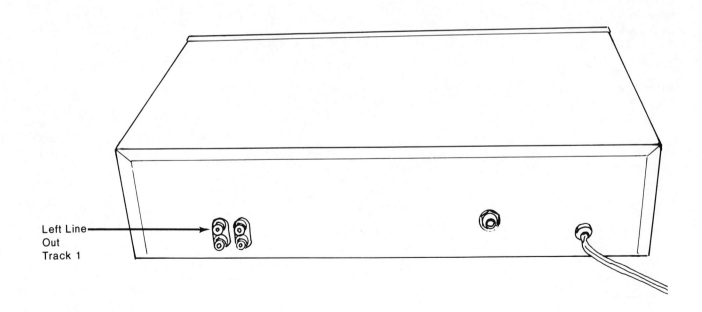

General Instructions

Setting Up

1. Assemble equipment and connect AV cart to AC outlet. Connect all equipment to power source.

2. Listen to audio on each track and determine which has the best audio. *Control tones* should be put on the track with the poorest audio.

Listening to Track 1 of Tape

3. Connect track 1 of tape deck to amplifier.

4. Separate signals of the two tracks.

5. Turn **on** tape deck *power control*.

6. Open cassette compartment and insert tape cassette.

Coyote, Tascam 225 Syncaset, Three Kodak Ektagraphic Slide Projectors, AVL Synchronizer (or Kodak Remote Control) and Amplifier Instructions

A multiple-outlet extension cord is a convenient way to connect the three projectors, Coyote, tape deck, and amplifier. Use the three 7-pin dissolve cables out of the Coyote to connect the projectors to the system. Each cable is identified as A, B, or C. A logical configuration would be A as the top projector, B as middle, and C as the bottom projector.

Control tones will be recorded on either track 1 or track 2, and controls and patch cords must be set for appropriate track.

Coyote Tascam 225, and Amplifier Instructions

Connect patch cord with RCA connector to *left line out jack* of tape deck and *phono stereo input jack* of amplifier.

Tascam 225 Syncaset Instructions

Turn *track 1 output pan* to far left and *track 2 output pan* to far right.

Depress *power control*.

Depress *eject control* and place tape cassette with prerecorded audio into deck.

Supplementary Information

The *115/220 control* underneath Coyote must be set to proper voltage.

The Tascam 225 Syncaset is capable of recording two parallel tracks. Audio will be left on one track and control tones will be put on the other track.

If output signals of each track are not separated, they will tend to spill over into the adjoining track and confuse the output function of deck.

Tape opening should be down and full reel of tape on the left.

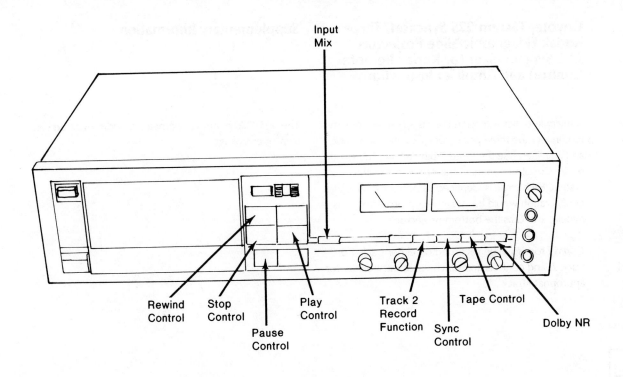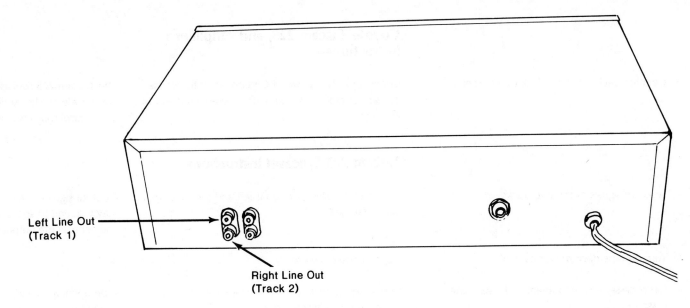

| **General Instructions** | **Amplifier Instructions** | **Supplementary Information** |

General Instructions

7. Turn **on** amplifier.

8. Rewind tape to beginning and listen to track 1.

9. Adjust volume.

10. Rewind tape and stop tape deck.

Listening to Track 2 of Prerecorded Tape

11. Connect track 2 of tape deck to amplifier.

12. Listen to track 2 (right track).

13. After determining which track has the best audio, erase the other track. As an example, we will assume that track 1 is best and will erase track 2, which will then receive control tones from the Coyote.

14. Prepare to erase track 2.

15. Erase track 2.

Amplifier Instructions

Turn *on/off control* to **on**. Volume will have to be adjusted when tape is running.

Amplifier and Tascam 225 Tape Deck Instructions

Depress *play control*.

Tascam 225 Syncaset Tape Deck Instructions

Switch patch cord from *left line out* to *right line out*.

Use labeled controls for *play, stop, rewind,* and *stop*.

Input mix control should be **off**. *Sync control* should be **off**. *Tape control* should be set on **normal** and, *Dolby NR control* should be **off**.

Depress *record function control* for track 2.

Depress *pause control*.

Depress *record* and *play* simultaneously. Release *pause control*.

Supplementary Information

Volume controls are not always located on *on/off control*.

Track 1 is left track and track 2 is identified as the right one.

A *red indicator light* will blink to indicate that the tape deck is ready to erase track 2.

Pause is used before and after starting *record functions* to eliminate a popping noise in recording. Pause may be also used when a delay is of short duration.

This will cause the *red indicator light* to stay on without blinking.

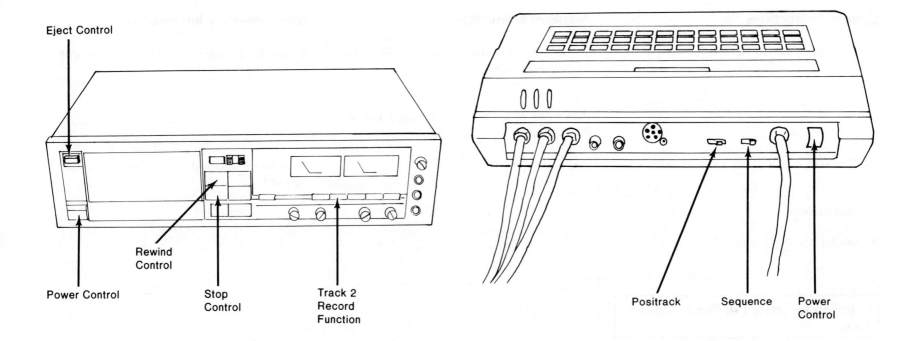
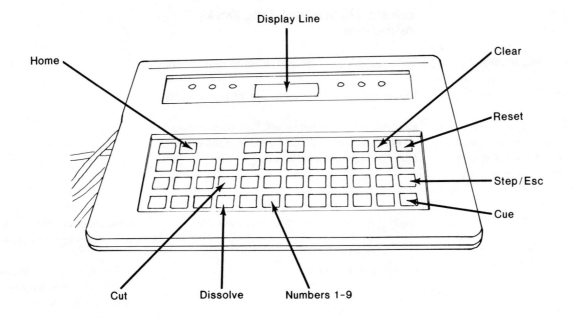

General Instructions

16. Cancel record ready mode.

17. Rewind tape.

18. Remove cassette.

Programming Coyote

19. Turn on *power control*.

20. Program Coyote.

Tascam 225 Syncaset Tape Deck Instructions

Depress *stop control* when track 2 has been erased.

Depress *record function control* for track 2. This will cancel record ready mode.

Activate *rewind control*.
Depress *stop control*.

Depress *eject control*. Remove cassette.

AVL Coyote Instructions

Positrak control should be **on**.
Sequence control should be on **3**. Depress *power control*.

Program Coyote with dissolve or cut commands only. Do not program any wait commands. If you want projectors to return home to beginning at end of program, program a home cue.

Press *reset* when finished if home cue was not used.

Supplementary Information

A storyboard with audio script, slide descriptions, and control tones anticipated is needed.

An example would be 2 Diss cue, 2 Diss cue, 3 Diss cue, and cut cue. Each command will show in the *display line* and will be numbered in consecutive order. Home is programmable. A 2 Diss cue is a two-second dissolve. A wait command is not needed. Time between slides is determined by operator. *Display line* will return to first cue.

Pressing *reset* will return *display line* to first cue, and all projectors will return home with lamps off. Reset is not programmable. Pressing *reset* and *clear* simultaneously will clear the *display line* and the memory of the Coyote.

Pressing *step/esc* will return the Coyote to its normal running mode and halt the current action that is being executed.

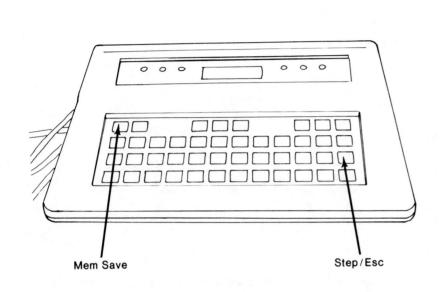
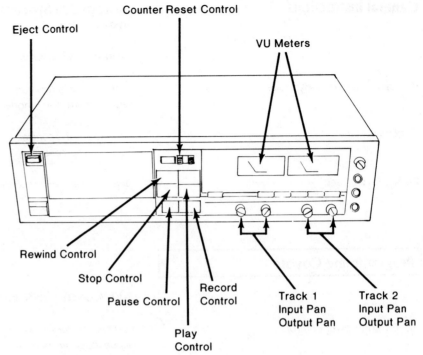
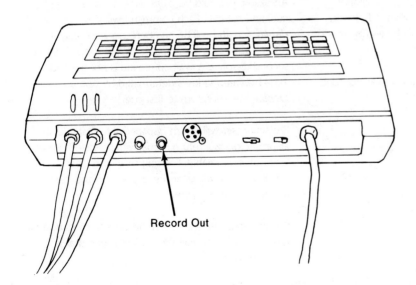
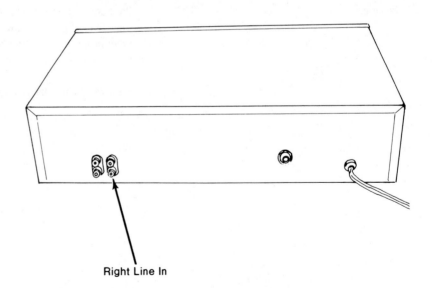

| **General Instructions** | **Tascam 225 and Coyote Instructions** | **Supplementary Information** |

Saving a Program in Coyote onto Tape

21. Connect Coyote to audio deck.

    Attach patch cord from *record out* on Coyote to *right line in* of Tascam 225 Syncaset.

    Parts of a program or a completed program should be saved on audiotape in case of power failure. On replay, it counts down rapidly.

22. Separate signals.

    Track 1 output pan should be turned to far left and *track 2 output pan* should be turned to far right.

23. Adjust volume of control signals that will be recorded.

    Depress *track 2 record function control.* Adjust *track 2 input pan* so that VU meter registers + 2 db or slightly more.

24. Insert audiocassette.

    Depress *eject control* and insert blank audiocassette. Depress *rewind control.* Depress *counter reset control.* Depress *pause control.*

    Counter reset should register zero.

25. Start recording.

    Depress *record* and *play controls* simultaneously.

    Depress and release pause *control.*

    Depress *mem save* on Coyote.

    Depress *step/esc* on Coyote.

    The Tascam 225 Syncaset must be running in record mode before depressing *memory save control* on the Coyote because the recording of the signals takes very little time. Make sure that the tape is well beyond the leader before introducing signals from the Coyote. You will hear *positrak* and be able to see the command number count down. At the end of the count, the program will return to its beginning.

26. Stop recording.

    Depress *pause control* when all commands are on tape. Depress *stop control.* Depress and release *pause control.*

69

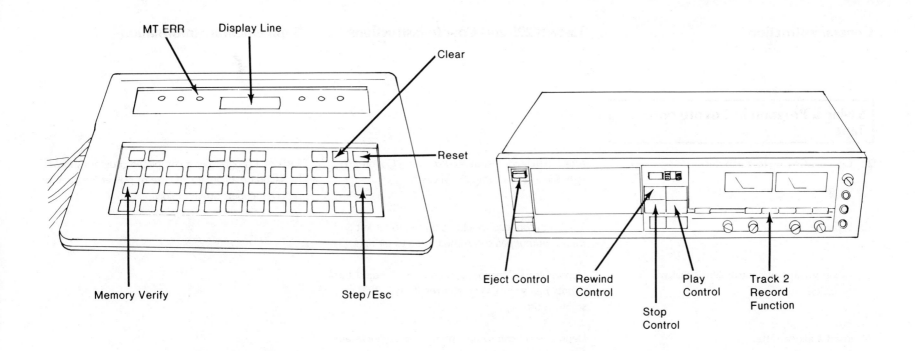
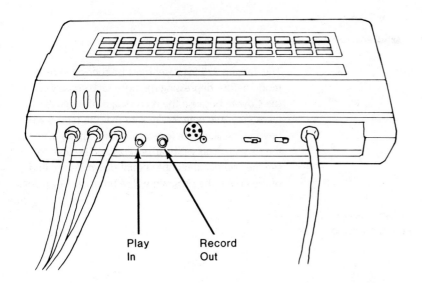
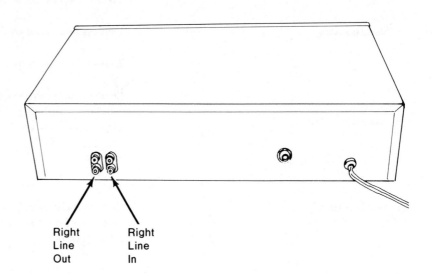

| General Instructions | Tascam 225 and Coyote Instructions | Supplementary Information |

27. Rewind tape.

Depress *rewind control*.

Depress *stop control*.

28. Cancel record ready mode.

Depress *record function control* for track 2.

Verifying if Coyote Program is on the Tape

29. Change patch cord.

Change patch cord that is connected to *right line in* of Syncaset 225 to *right line out* and change the connections on Coyote from *record out* to *play in*.

30. Verify program on tape.

Depress *memory verify* on Coyote.

Depress *play* on Syncaset 225.

When Tascam 225 Syncaset counter reaches 5, you will be able to see command numbers count down on the Coyote. Program will return to beginning in *display line*. If *display line* exhibits BAD and M + Err lights on the Coyote, there is an error in program and you must depress *clear* and *reset* on Coyote in order to reprogram it.

31. Stop Tascam 225 Syncaset recorder.

Depress *step/esc* on Coyote.

Depress *stop control* on Tascam 225.

Depress *rewind control* on Tascam 225.

Depress *stop control* on Tascam 225.

32. Remove tape.

Depress *eject control* and remove tape cassette.

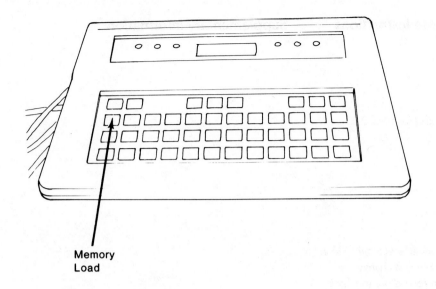
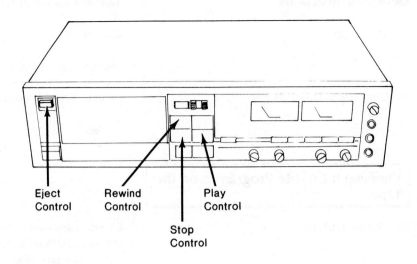
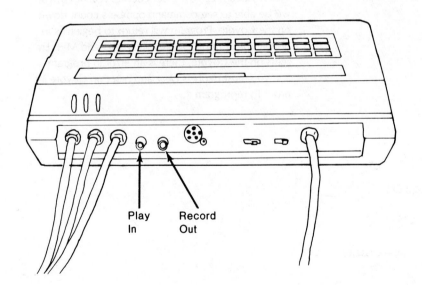
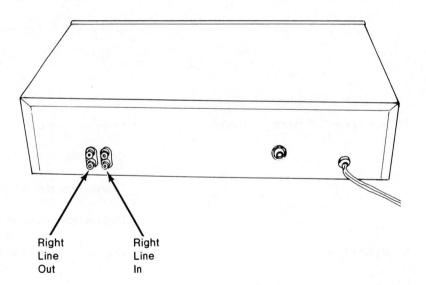

| **General Instructions** | **Tascam 225 and Coyote Instructions** | **Supplementary Information** |

Loading a Program on Tape into Coyote for Presentation

33. Insert tape.

Insert tape with control tones saved from Coyote program.

This procedure is necessary if the program in the Coyote is lost. It will restore the information for the program in the Coyote.

34. Change patch cords.

Connect patch cord to *right line out* of Tascam 225 and *play in* of Coyote.

35. Rewind tape, if necessary.

Depress *rewind control* on Tascam 225.

Depress *stop control* on Tascam 225.

36. Load program into Coyote.

Depress *memory load* on Coyote.

Depress *play* on Tascam 225.

Command numbers will count down in display window of Coyote.

37. Stop Tascam 225.

Depress *stop control* on Tascam 225 when numbers have stopped counting down.

38. Rewind tape.

Depress *rewind control* on Tascam 225.

Depress *stop control* on Tascam 225.

39. Remove cassette.

Depress *eject control* and remove cassette, which now stores Coyote program.

Marrying Audio to Coyote Program

40. Change patch cord.

Connect patch cord from *right line in* of Tascam 225 and *record out* of Coyote.

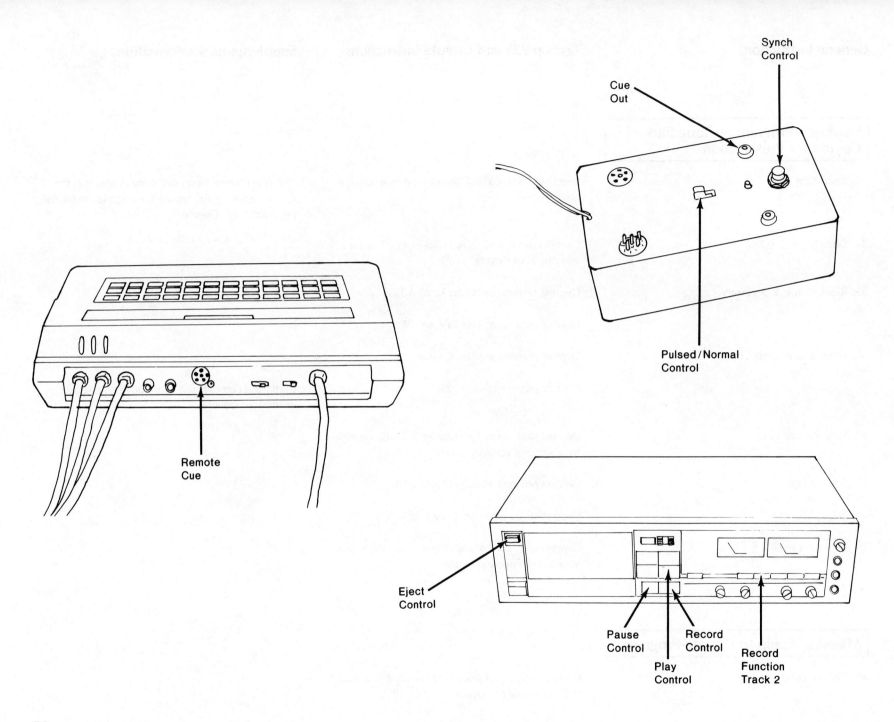

General Instructions	**Tascam 225, Coyote, and AVL Synchronizer Instructions**	**Supplementary Information**
41. Connect Coyote and AVL Synchronizer.	Connect patch cord to *remote cue* of Coyote and *cue out* of AVL Synchronizer (or connect Kodak *remote control* to *remote cue* of Coyote).	If the AVL Synchronizer is used, it will be connected to a power source. *Pulsed/normal switch control* on the AVL Synchronizer should be in the pulsed position.
42. Insert cassette with audio.	Depress *eject control* on Tascam 225 Syncaset and insert cassette that has audio on track 1 and a blank track 2.	
43. Prepare to record.	Depress *record function control* for track 2.	
44. Turn **on** amplifier.		

Slide Projectors Instructions

45. Place slide trays on projectors.	Trays should be properly seated and advanced to number 1.	
46. Turn **on** projectors.	Turn each projector to **fan.**	Turning to lamp would bypass the *on/off function* of the program from the Coyote.
47. Focus and align projectors.	Adjustment screws are available on projectors and stand to provide lateral and vertical movement.	Projectors may be turned to lamp one at a time to check size and border alignment and then all projected to determine that images overlay. Remember to turn back to **fan.**
48. Start recording.	Depress *record control* and *play control* simultaneously.	

AVL Synchronizer or Kodak Remote

49. Activate AVL or Remote Control.	Depress *sync control* on AVL Synchronizer or Kodak Remote Control each time a different slide is desired.	Commands in *display window* of Coyote will advance to the next command each time the *sync control* is depressed.

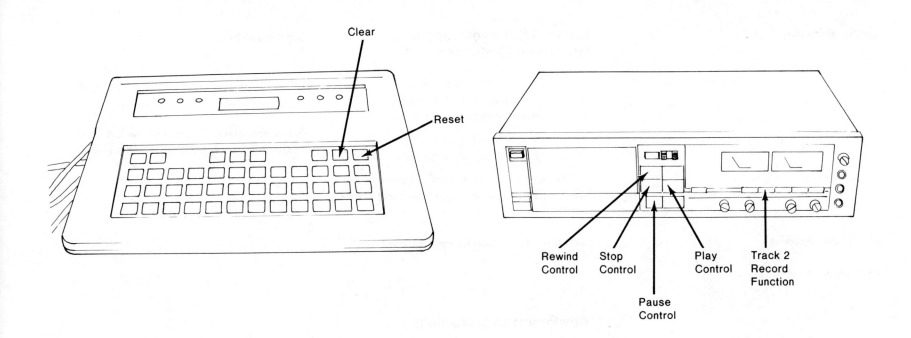
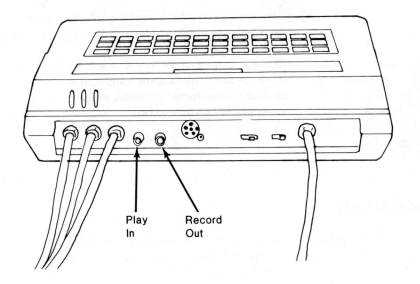
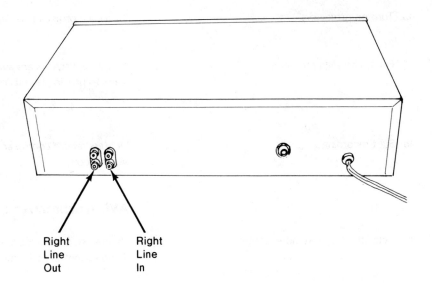

General Instructions

50. Stop recording.

51. Cancel *record function*.

52. Rewind tape.

Tascam 225 Syncaset Instructions

Depress *pause control* when show has been synchronized.

Depress *stop control*.

Release *pause control*.

Depress *track 2 record function control*.

Depress *rewind control*.

Depress *stop control*.

Supplementary Information

Playing Back the Complete Program

Tascam 225 Syncaset and Coyote Instructions

53. Clear Coyote memory.

54. Change patch cords.

55. Check slide tray positions.

56. Start program.

Depress *reset* and *clear* simultaneously on Coyote.

Change patch cord from *right line in* to *right line out* on Tascam 225 and *record out* to *play in* on coyote.

If necessary, reset each tray so that the number 1 slot is opposite arrow on projectors.

Depress *play* on Tascam 225 and enjoy the show.

This will clear the memory of Coyote.

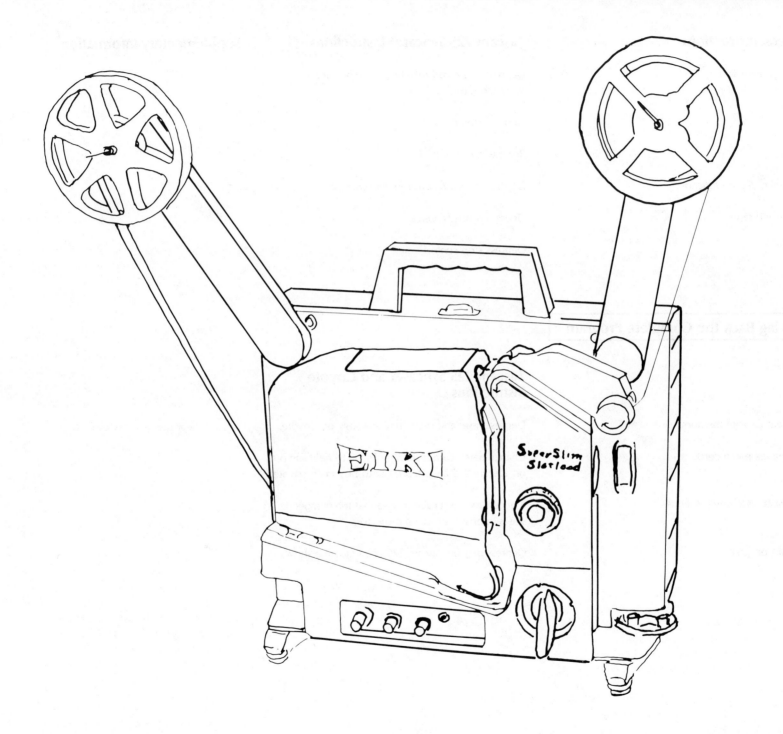

Slot-load 16mm Sound Motion Picture Projector

One of the most frequently used pieces of AV equipment is the 16mm motion picture projector. Research has provided abundant evidence of the value and effectiveness of films in classroom instruction. Extensive libraries of educational films are accessible to most teachers. Essential information necessary to the operation of a 16mm projector (including a diagram of the threading procedures) is usually found imprinted on the machine or on the inside of the cover. Automatic loading projectors are becoming increasingly popular because they appear less complex to the novice user. If all goes well, projecting a film may be accomplished in a few simple steps.

Objectives

Observable Behavior

1. Set up projector and show a portion of a 16mm sound film.

2. Rewind film and return to reel container.

3. Restore projector to storage conformation.

Conditions

Given a correctly wound 16mm sound motion picture film. Time limit 3 minutes.

Time limit 2 minutes.

Time limit 1 minute.

Level of Acceptable Performance

Correctly projected film with picture in focus, screen width fully utilized, and crisp, clear sound synchronized with picture.

Film must be correctly rewound.

Cords stowed and case closed.

Essential Parts

Typical

Cover
Film channel
Lamp housing
Lens
Power cord
Reel arms
Spindle
Supply reel
Take-up reel

Controls:
  Treble
  Bass
  Amplifier on/off and
    volume
  Remote jack (ESL)
  Inching knob
  Focus
  Framing lever
  Elevator
  Function (motor/lamp)
  Still picture clutch

Additional on Eiki

Antitheft lens screwlock
Black heat shield
Exciter lamp
Film guard
Inching knob
Lamp ejection lever
Cover
Handle
Latch
Power cord storage
  compartment

Controls:
  Hi-Low lamp
  Opt/mag
  Still/run
  Reel arm lock

Available on Some Models

Push-button function selector

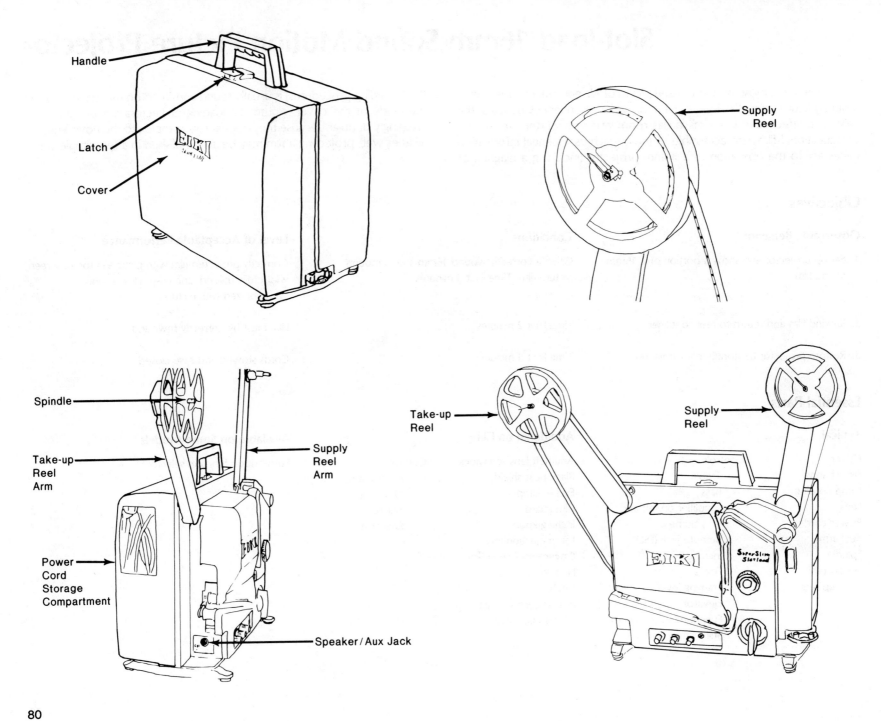

| General Information | Eiki Instructions | Supplementary Information |

Threading

1. Read Projection Principles, Audio Principles, and AV Equipment Principles before proceeding.

2. Position projector on projection stand or cart so that it faces screen.

3. Remove cover from unit.

   Lift latch under handle.

4. Tie *power cord* around bottom of projection stand or cart and attach to grounded power outlet.

   Power cord is stored in compartment at left rear of projector.

5. Connect and position speaker if it is not built in.

   A speaker is built into the rear cover. The *speaker/aux jack* is at rear of machine.

   Some projectors have a separate speaker which may be positioned a distance from projector. It should be placed near screen, at ear level, facing audience.

6. Position *reel arms*.

   Take-up reel arm should be moved from horizontal to vertical until it locks into place. *Supply reel arm* should be rotated until it clicks into a nearly vertical position.

   Most modern projectors have permanently mounted reel arms that can be swung into position; this action sometimes requires operator to press a *reel arm release control*.

7. Place *take-up reel* on *take-up reel spindle*.

   Square-hole side on reel must be placed on shaft first. A spring-loaded retainer on spindle will keep reel from falling off.

   Whenever possible, *take-up reel* should be same size as *supply reel* because many film libraries suggest that films be returned without rewinding. Never use *take-up reel* smaller than *supply reel*.

8. Place *supply reel* on *supply reel spindle*.

   Film should fall from reel in a clockwise direction with sprocket holes toward operator.

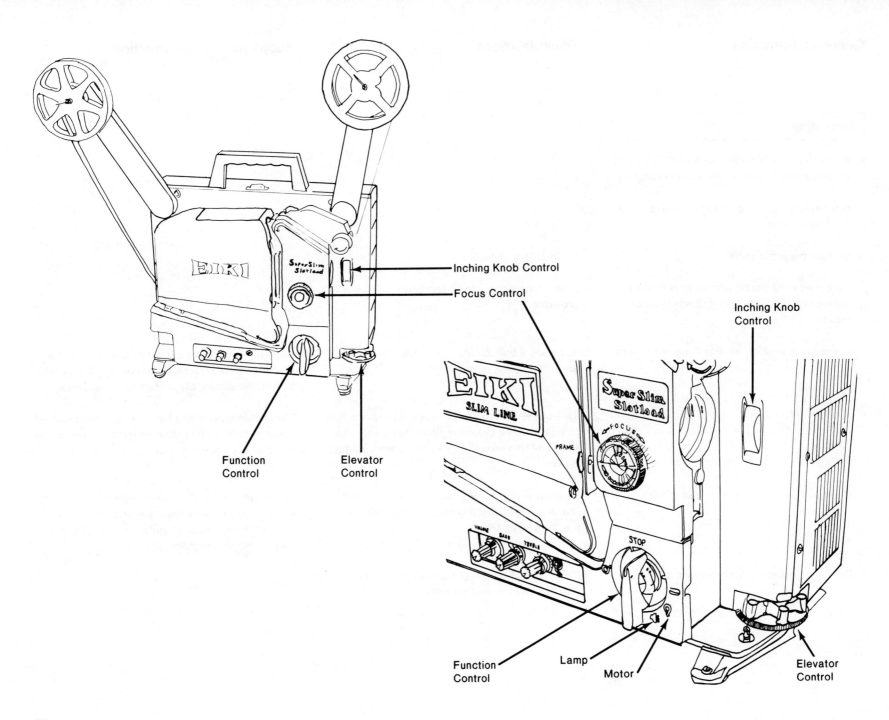

| General Instructions | Eiki Instructions | Supplementary Information |

Prefocusing

9. Turn motor and lamp **on**.

Move *function control* to **forward** and **lamp** position.

On most machines it is not possible to turn on lamp without motor because cooling fan is driven by same motor that drives transport system.

10. Adjust projected light pattern to insure proper alignment, image size, and preliminary focus.

Rotate *focus control knob*. Adjust *elevator control* to raise or lower projector.

Prefocusing eliminates need to run film through projector to determine adequacy of projection size and alignment. Adjust lens to produce clean, sharp edges. Image size is increased by moving projector farther away from screen.

11. Turn motor and lamp **off**.

Move *function control* to **stop**.

Some machines are equipped with push-button controls.

Loading

12. Set automatic threading mechanisms.

Turn *function control* to **stop** position.

Some projectors may have **off** position.

13. Thread film.

Film should be pushed lengthwise into indicated loading path. Do not thread film like manual or self-thread machines require.

14. Attach leader end to *take-up reel* and rotate reel clockwise to take up film slack.

The *inching knob control* may be rotated manually to advance film to determine if it has been threaded correctly.

Function control knob must be at approximately 3 o'clock position for *inching knob control* to manually advance film.

Making Preshow Adjustments

15. Start projector.

Turn *function control* to **lamp** position.

16. Refocus.

Rotate *focus control* to produce a sharp image.

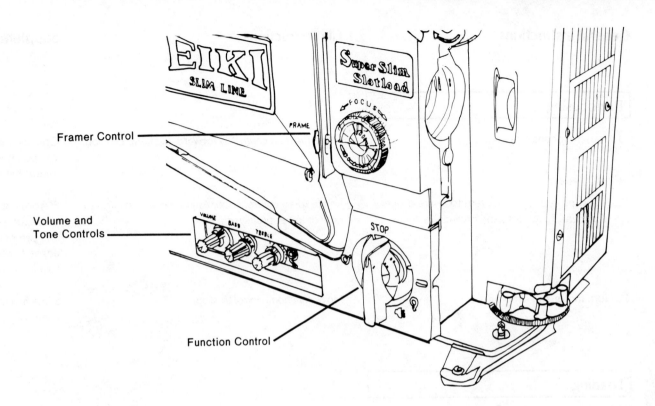
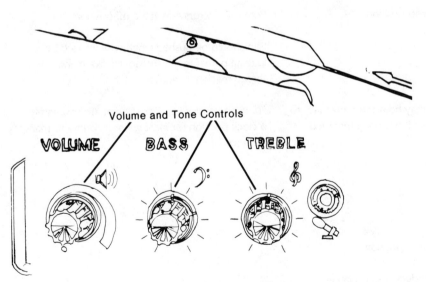

General Instructions	Eiki Instructions	Supplementary Information
17. Adjust *volume* and *tone controls*.	Rotate amplifier *on/off volume control*. Adjust *bass* and *treble controls*.	Some models have a *voice/music switch* with a single *tone control knob*.
18. Turn projector **off**.	Move *function control* to **stop**.	
19. Reverse film back to beginning of vital content.	Move *function control* counterclockwise to **off** and then on to **reverse** position.	

Showing a Film

General Instructions	Eiki Instructions	Supplementary Information
20. Turn volume to **low**.		
21. Turn lamp **on** and increase volume slowly.	Move *function control* to **lamp**. Rotate *volume control*.	Operator should move to farthest corner of room to determine quality and volume of sound. It will be quite different with a roomful of people than when the room is empty.
22. Adjust *framer*.	Move *framing lever control* up or down to eliminate visible bar in projected image.	
23. It may be useful to stop film and back it up to repeat a scene.	Move *function control* to **off** and then counterclockwise to **reverse** position. Amplifier *on/off volume control* should be turned down.	Garbled sound as film travels backward can be annoying to an audience.
24. At conclusion of film, gradually turn down volume, turn lamp off after last credits, and turn motor off after trailer has run through machine.	Rotate *function control* to **off**.	Trailer is usually opaque material that contains no information. It provides a convenient means of handling film without touching fragile information portions.

NOTE: If motion picture film should break during projection, stop machine and go through all threading operations again. When film has traveled through projector, place broken end of film under broken end of film on take-up reel. Turn take-up reel by hand several turns to cause pressure of film layers to hold broken ends in place. Place strip of paper through reel at point of breakage as it winds on reel to indicate point of break.

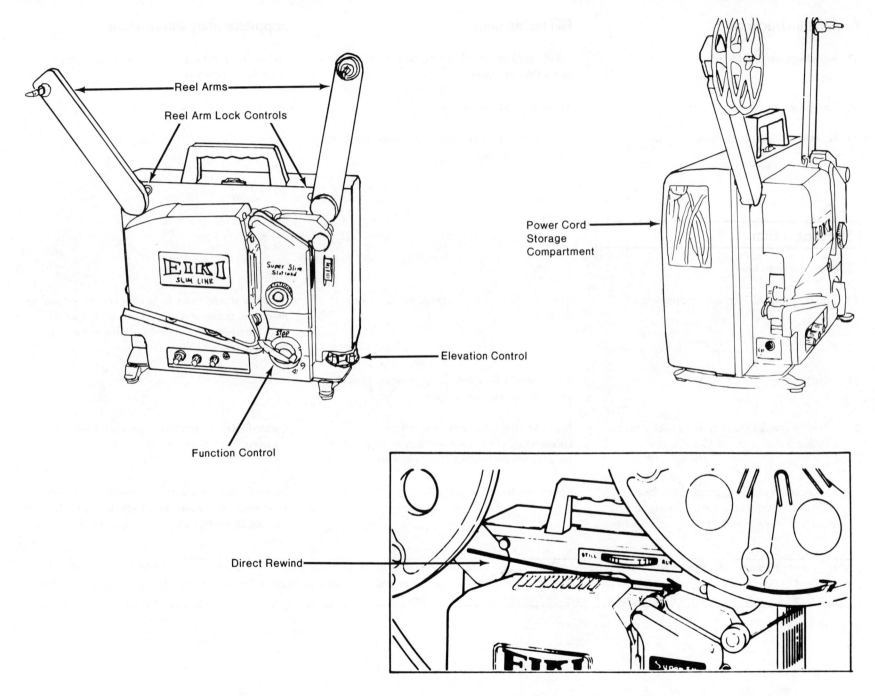

| **General Instructions** | **Eiki Instructions** | **Supplementary Information** |

Rewinding

25. Rewind film.

For in-path rewind, stop projector before end of film trailer and move *function control* to **rewind.** Do not turn *function control* to **stop** until all of the film is on supply reel.

For direct rewind, run film completely through machine. Bring end of film trailer to empty supply reel. Pass film under supply reel hub and insert in slot of hub.

26. Remove *supply* and *take-up reels*. Return film to film container.

27. Lower reel arms after depressing *supply arm lock controls*.

28. Make certain that *function control* is on **off** or **stop.**

Rotate *function control*.

Function control must be left in an inactive status to prevent damage to film contact rollers in transport system. Some models have push-button controls.

29. Lower *elevation control*.

30. Disconnect *power cord* and store in compartment provided.

87

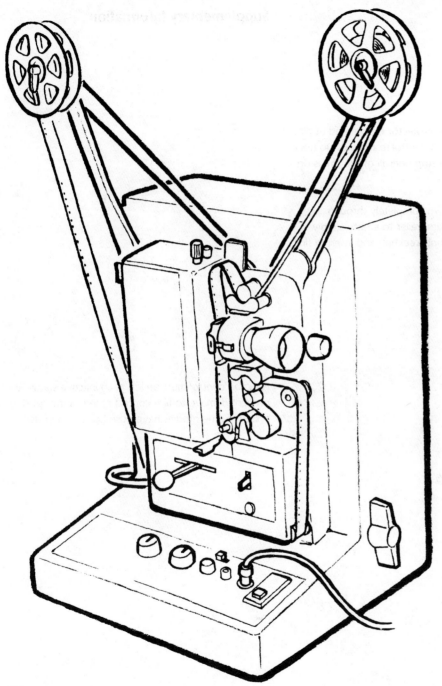
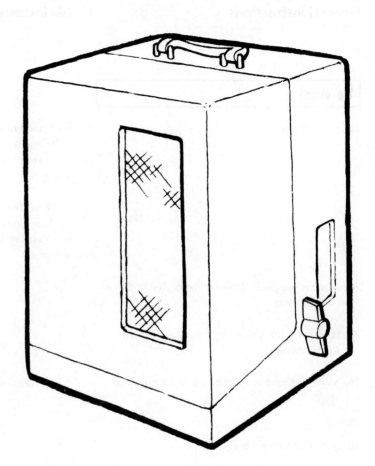

PROJECTOR ENCASED

Manual-thread 16mm Sound Motion Picture Projector

One of the most popular pieces of AV hardware is the 16mm sound motion picture projector. Research has provided substantial evidence of the effectiveness of films in instruction, and large libraries of educational films are accessible to most teachers. It is capable of producing sound and motion and, therefore, provides the most realistic reproduction of reality. Essential information necessary to the operation of a 16mm sound projector (including a diagram of the threading path) is usually imprinted on most machines.

Objectives

Observable Behavior

1. Set up projector and show a portion of a 16mm sound film.

2. Rewind film and return it to reel container.

3. Restore projector to storage conformation.

Conditions

Given a correctly wound 16mm sound motion picture film. Time limit 3 minutes.

Time limit 2 minutes.

Time limit 1 minute.

Level of Acceptable Performance

Correctly formed upper and lower loops; picture in focus; screen fully utilized; crisp, clear audio; sound and picture synchronized.

Film must be correctly wound.

Cords stowed and case closed.

Essential Parts

Typical

Film channel
Lens
Lens housing
Lower drive sprocket
Power cord
Reel arms
Speaker
Sound drum
Spindle
Supply reel
Take-up reel
Upper drive sprocket
Controls:
    Elevation   Lamp
    Framing   Motor

Additional on Kodak Pageant

Belt
Exciter lamp indicator
Loop restorer
Pressure plate
Reel retainers
Speaker jack
Speaker plug
Controls:
    Amplifier   Tone
    Fidelity   Volume
    Functions (motor, lamps)
    Manual transport
    Reel arm release
    Rewind
    Sound input

Available on Some Models

Controls:
    Automatic safety clutches
    Magnetic audio record
    Still frame

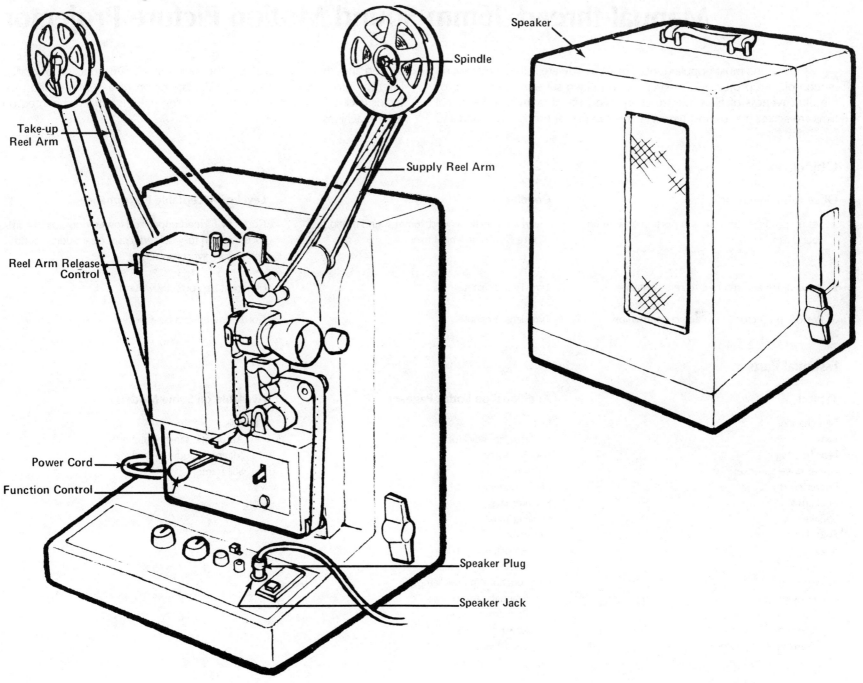

| General Instructions | Kodak Pageant Instructions | Supplementary Information |

Threading

1. Read Projection Principles, Audio Principles, and AV Equipment Principles before proceeding.

2. Position projector on projection stand so that it faces screen.

3. Open projector case.

    Raise *take-up reel arm* to facilitate *power cord* removal.

4. Tie *power cord* around bottom of a projection stand leg.

5. Turn motor and lamp **off.**

    Move *function control* to **off.**

6. Plug *power cord* into electrical outlet.

7. Correct and position *speaker*.

    Remove speaker cord from inside cover and connect *speaker plug* to *speaker jack*.

    Some *speakers* are integrated into case and need not be connected manually. If speaker is separate from projector, place it near the screen, **off the floor,** facing the audience. If there is danger of someone tripping on cord, it is advisable to place *speaker* near projector.

8. Position *take-up reel arm*.

    Take-up reel arm is rotated upward until it clicks into a holding position.

    Normally *supply reel arms* are located at top and front of projector. Kalart/Victor projectors use rear top position. Most modern projectors have permanently mounted *reel arms* that can be swung into position; this action sometimes requires operator to press a *reel arm release control*. If *reel arms* are not permanently mounted, they must be positioned with *spindle* pointing to the right.

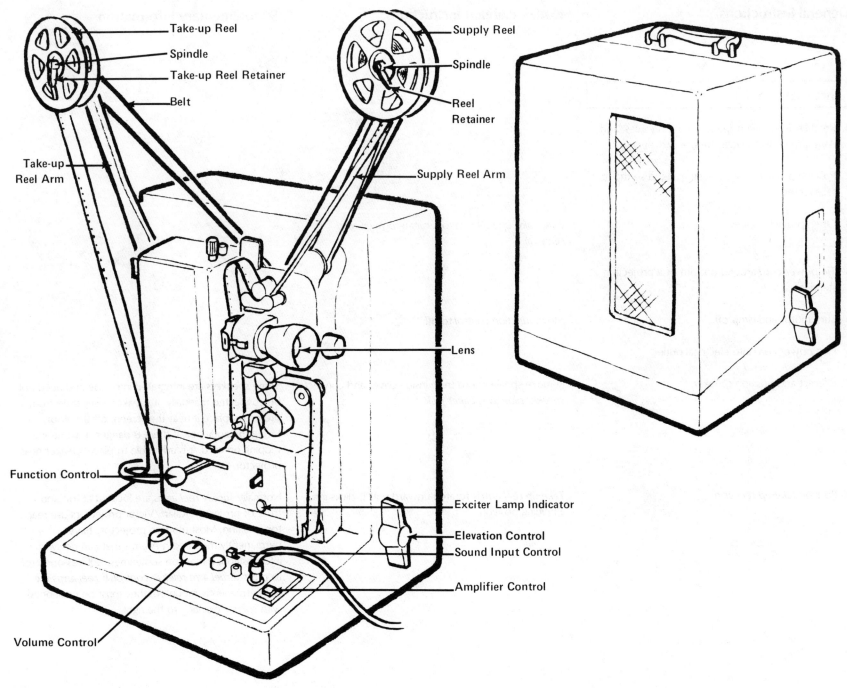

General Instructions	Kodak Pageant Instructions	Supplementary Information
9. Position *supply reel arm*.	*Supply reel arm* is rotated forward until it stops.	
10. Position *take-up reel arm*.	Ropelike *belt* must properly engage pulley on *take-up reel arm*.	Springlike *belts* must sometimes be pulled from the machine and engaged on *reel arm* pulleys. Some *belts* are enclosed within *reel arm* and require no action by user.
11. Place empty *take-up reel* on *spindle*.	Secure reel by rotating *take-up reel retainer* on the *spindle* 90°.	Some *spindles* use spring-loaded *reel retainers* which require only that reel is pushed far enough on shaft to engage it. Whenever possible, *take-up reel* should be same size as *supply reel* because many film libraries suggest that films be returned without rewinding them. Never use a *take-up reel* that is smaller than *feed reel*.
12. Turn motor **on** and observe movement of *take-up reel*. If it does not turn clockwise, drive *belts* are probably twisted.	Move *function control* to **forward-rewind/motor**.	
13. Turn *amplifier* or *on/off control* **on.**	Depress *amplifier control,* which should cause *exciter lamp indicator* to glow. Move *sound input control* to **film. Mike** position allows user to substitute own commentary as film is shown.	*Amplifier controls* are often combined with *volume controls*. Switching *amplifier control* **on** during threading allows time for warmup of tube-type amplifiers.
14. Turn motor and lamp **on.**	Move *function control* to **forward-rewind lamp**.	On most machines it is not possible to turn on lamp without motor because cooling fan is driven by same motor that drives transport system.
15. Adjust projected light pattern to insure proper alignment, image size, and preliminary focus.	Rotate *lens* barrel; turn *elevation control* if required.	Prefocus the projector to determine adequacy of projection size and alignment. Adjust projection *lens* to produce clean, sharp edges. Adjust *elevation control* to raise or lower projected image. Image size is increased by moving projector farther away from screen.
16. Turn motor and lamp **off.**	Move *function control* to **off**.	
17. Place *supply reel* on *supply spindle*.	Secure *reel* with *reel retainer*.	Film should leave *supply reel* in clockwise direction with sprocket holes toward operator; image should be upside down. Any other condition indicates that film was not properly rewound after last showing.

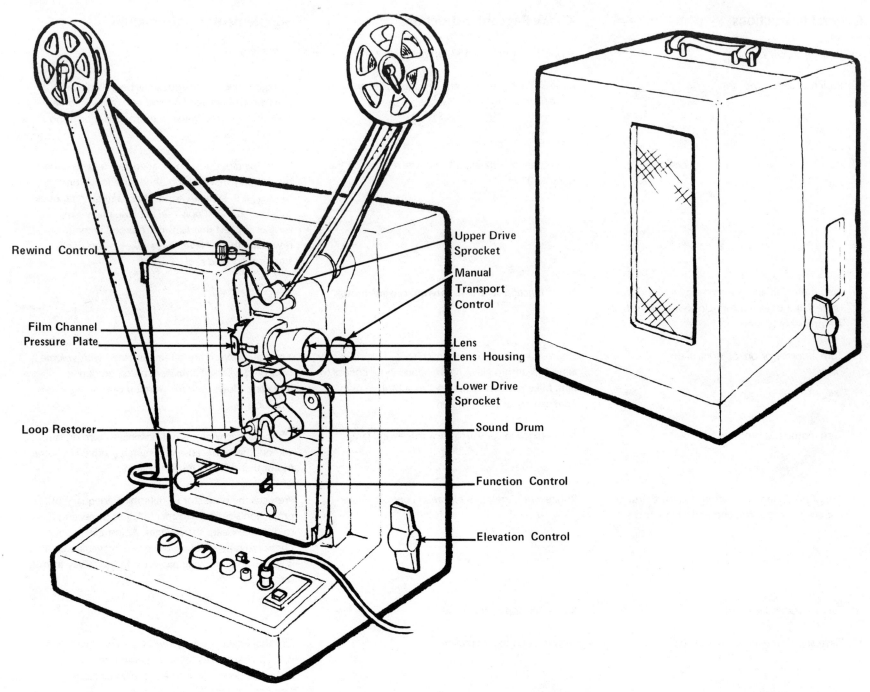

General Instructions	**Kodak Pageant Instructions**	**Supplementary Information**
18. Open retainers around *upper* and *lower drive sprockets* to facilitate threading.	Move clamps out from *drive sprockets* toward *lens housing*.	Typically, film retaining devices must be moved out of operating position to facilitate threading film. Projectors usually have at least two *drive sprockets;* one above the *film channel* and one below the *film channel* near the *sound drum*.
19. Open *film channel*.	Push *pressure plate* forward until it is caught by spring clip on *lens housing*.	On some machines, the *lens housing* is swung open like a door on a hinge and is located at rear of the housing. Others have a lever located at rear of *lens housing* that opens *film channel*. The *pressure plate* must be moved to allow film to be positioned in *film channel*.
20. Pull about five feet of film leader from *supply reel*.	Check *rewind control* position; it should be **upright**.	Do not let film touch floor because dirt will lodge in *film channel* and scratch film as it passes through.
21. Rotate *manual transport control* to retract the claw behind the surface of the *film channel*.	Turn *manual transport control* until **white line** is toward you.	Moving film up and down in channel ensures that it is properly seated.
22. Position film to conform to threading diagram on projector.	Depress *loop restorer* to determine that lower loop is correct size. Loss of lower loop may cause picture to flutter, flicker, jump, or blur. To reform loop, depress *loop restorer* while projector is running.	The purpose of the film loops on a threaded 16mm sound projector is to provide for intermittent advance of film through channel and act as shock absorbers to prevent film breakage. Loop size must conform to pattern on machine. If lower loop is too long or short, sound and picture will be out of synchronization, because sound is exactly 26 frames ahead of the picture. The *photoelectric cell* in a sound motion picture projector responds to projected beam from *exciter lamp* as its beam is passed through film with optical sound track. Variations in density of track are read by *photoelectric cell* and are transduced to electrical energy on way to *amplifier*. Most 16mm sound film has an optical sound track. This runs the length of film opposite sprocket holes. Sound from *speaker* will seem garbled if film is not snug around *sound drum*. Generally, tension rollers in this area maintain proper tension if threaded properly.

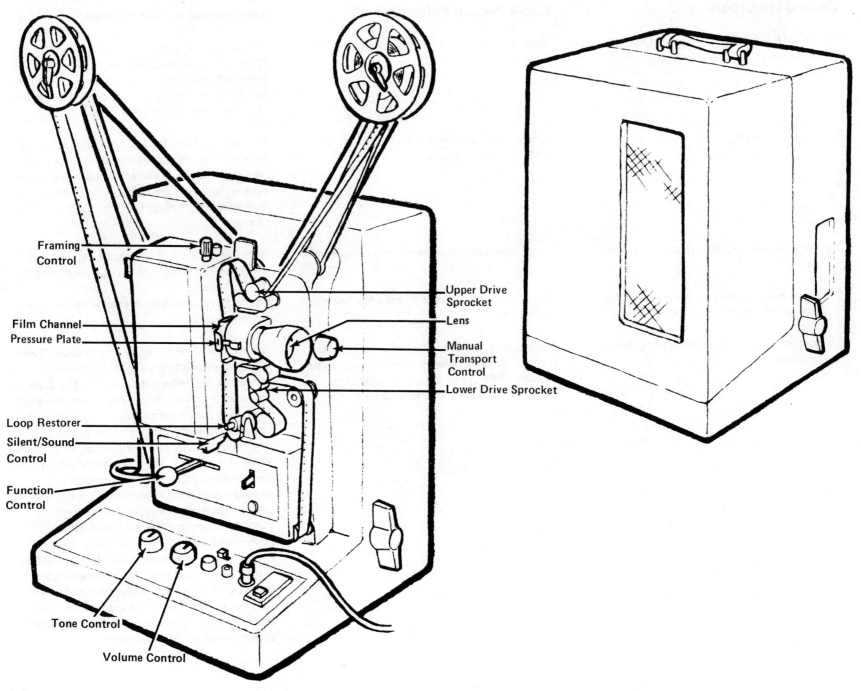

| **General Instructions** | **Kodak Pageant Instructions** | **Supplementary Information** |

23. Close *film channel*.

Depress *spring clip* and *pressure plate* will snap *film channel* closed.

24. Check film to determine that *drive sprocket teeth* are engaged in film perforations.

25. Close *upper* and *lower drive sprocket* clamps.

Close clamps by moving them away from *lens housing*.

26. Rotate *manual transport control* and observe travel of film through projector.

Showing a Film

27. Turn volume **low**.

It is recommended that a film be started with sound at a low level and brought up to a pleasing volume gradually.

28. Turn motor **on** and run film through the machine until the beginning of film may be observed approaching *film channel*.

Move *function control* to **forward-rewind.**

It is suggested that audience be spared leader and visual identification material before title appears.

29. Turn lamp **on**.

Move *function control* to **forward-rewind/lamp.**

30. Focus image.

Rotate *lens* barrel until image is clear and sharp.

31. Check *silent/sound control* to determine that it is on **sound**.

The control may be moved only when projector motor is running.

This control regulates the speed with which film travels through projector. Silent speed is 16 frames per second and sound speed is 24 frames per second. Silent speed, if used with sound film, will produce a slight flicker on screen and audio will produce low-pitched, gutteral sounds.

32. Adjust *framing control* if parts of two frames appear on screen.

Framing control is located on top of projection bulb housing.

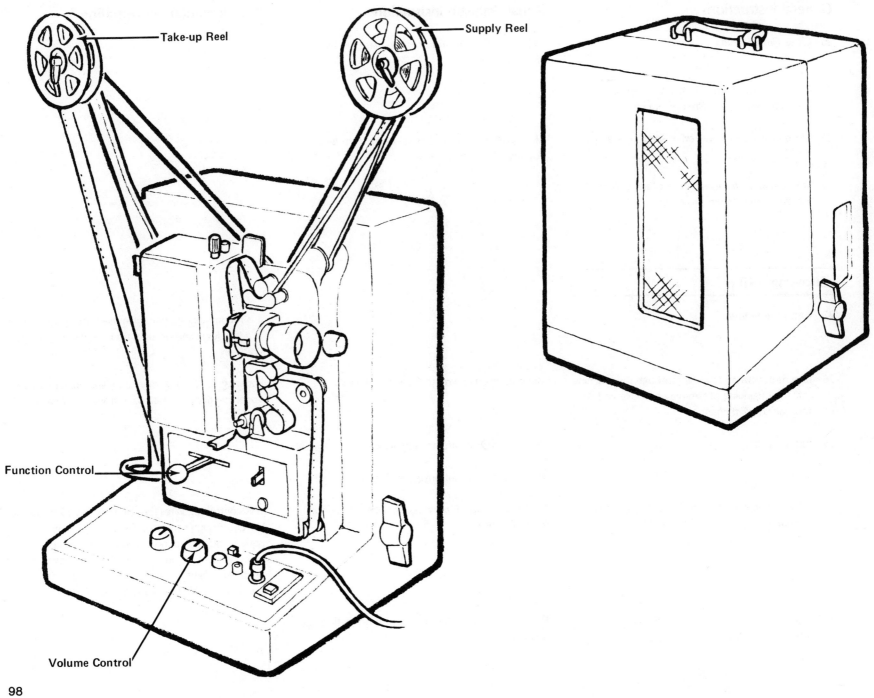

| **General Instructions** | **Kodak Pageant Instructions** | **Supplementary Information** |

33. Adjust audio controls to satisfactory level and tone balance.

Turn *volume control* clockwise to appropriate level. Adjust *tone control* to obtain satisfactory treble/bass balance. Position *fidelity control* to obtain crisp and clear audio; this position will vary from film to film.

Move to farthest corner of room to determine whether volume, tone, and fidelity are satisfactory.

34. It may be useful to stop film and back it up to repeat a scene.

To repeat a scene, move *function control* to **reverse/lamp,** pausing momentarily at **off** to allow transport system to stop forward motion.

Turn volume to a low level so that audio track running backward through machine does not disturb audience. Some projectors do not have reverse capability.

35. When desired point has been reached, turn machine **off** and readjust audio levels. Turn motor and lamp **on.**

Move *function control* to **off.** Pause a moment before moving to **forward-rewind/lamp** to begin showing film again. This is necessary to give transport system time to change direction and should be rigidly observed to prevent damage to projector.

Some projectors also provide single-frame or still-picture capability which enables the audience to view one frame for a period of time.

36. At conclusion of film, gradually turn down volume, turn lamp **off** as last credits fade, and turn motor **off** after trailer has run completely through the machine.

Turn *volume control* counterclockwise to lower audio level, move *function control* to **forward-rewind/motor** and then to **off.**

Turn projector lamp and audio **off** at close of information portion of film. The trailer is usually opaque material that contains no information. It provides a convenient means of handling film without touching the sensitive information portions.

If motion picture film should break during projection, place broken end of film under broken end on *take-up reel,* then turn *take-up reel* by hand several turns to cause pressure of film layers to hold broken ends in place. Place strip of paper at right angles to film at point of break as it winds on reel to indicate that film needs repair.

Rewinding

37. Attach trailer of film to *supply reel.*

Bring end of film trailer to empty *supply reel.* Do not attempt to run it back through projector. Pass film under the *supply reel* hub and insert it into slot in hub.

Some projectors require transfer of filled reel and empty reel to the opposite spindles before rewinding.

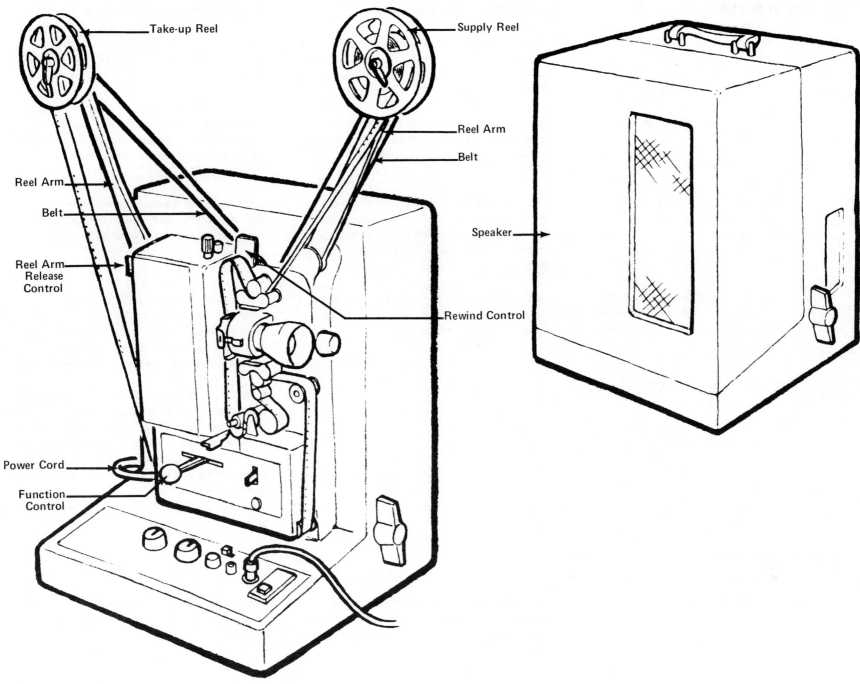

General Instructions	Kodak Pageant Instructions	Supplementary Information

38. Rewind the film.

Pull the *rewind control* downward to a nearly horizontal position. Move *function control* to **forward-rewind/motor.** Do not use **reverse.**

Rewind procedures will vary depending on make and model but most *rewind controls* are clearly labeled on the machine. The film should move rapidly from full *take-up reel* to empty *supply reel*. If film does not rewind rapidly, it is possible that *rewind control* or other appropriate control was not fully activated.

39. When all of the film has been transferred back to *supply reel,* turn machine **off.** Return *rewind control* to operating mode.

The *rewind control* should be returned to a vertical position.

On many projectors, the basic difference between **reverse** and **rewind** positions is that **reverse** is a projection mode. Film is not threaded through projector during **rewind**. **Rewind** affects travel of film as it moves from take-up to supply reels separate from transport system.

Stowing

40. Remove *supply* and *take-up reels*. Return film to film can.

Some projector cases or covers provide storage space for a *take-up reel.*

41. Disconnect projector and return *power cord* to storage compartment.

42. Restore *reel arms* to storage conformation.

Fold *supply reel arm* to storage position with *supply reel belt* left intact. Remove drive *belt* from *take-up reel* shaft. Apply pressure from left to right of *reel arm release control* to allow *take-up arm* to be lowered.

On some machines it may be necessary to disconnect *belts* and unbolt or otherwise detach *reel arms* from the projector.

43. Disconnect *speaker* and restore cord to storage.

Wind cord around retainers inside *speaker case.*

44. Replace cover on projector.

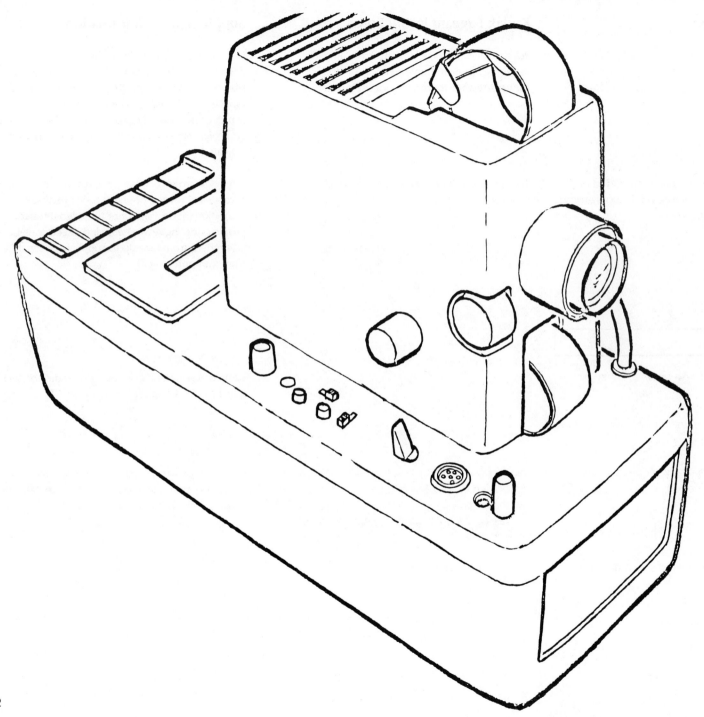

Filmstrip Projector with Audiocassette Playback

The obvious advantage of having both the projector and the audio playback for classroom presentations makes the sound filmstrip projector a valuable tool for large or small presentations. It is compact, easy to handle, and versatile. Filmstrips are easily stored and keep the visuals in a permanent sequential order. They are less expensive than slides or the 10 X 10 inch overhead transparencies. The automatic advance feature insures coordination of the visual and audio components.

Objectives

Observable Behavior

1. Set up projector to show a filmstrip with synched audiocassette.

2. Rewind filmstrip and return to container. Rewind audiocassette and remove from machine.

3. Projector restored to storage conformation.

Conditions

Given a correctly wound filmstrip and cassette. Time limit 3 minutes.

Time limit 2 minutes.

Time limit 30 seconds.

Level of Acceptable Performance

Filmstrip correctly threaded to move through machine. Properly framed, focused, and adjusted to fill screen area. Volume and tone controls appropriately adjusted.

Filmstrip to be rewound with tail inside, small enough to fit into can without "cinching." Audiocassette rewound to beginning of program.

Projector leveled, lens retracted, power cord retracted, and lid closed.

Essential Parts

Typical

Audiocassette controls:
    Fast forward
    Pause
    Play
    Rewind
    Stop/eject

Projector controls:
    Advance (ADV)
    Elevation
    Film speed
    Focus
    Framer
    Manual/auto
    Off/fan/lamp
    Volume

Additional on Dukane

Remote control

Available on Some Models

2 X 2 slide stack holder

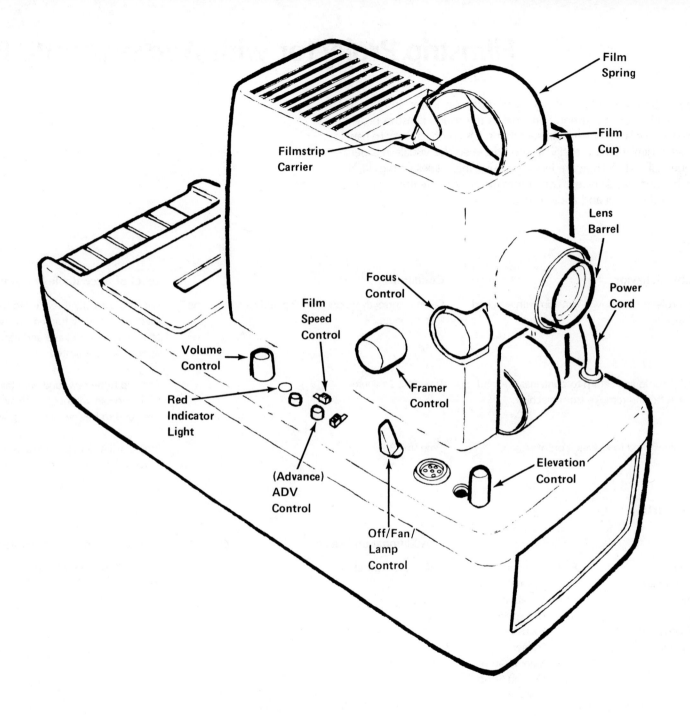

| **General Instructions** | **Dukane Micromatic II Sound Filmstrip Projector Instructions** | **Supplemental Information** |

Projecting a Filmstrip

1. Read Projection Principles and Audio Principles before proceeding.

2. Place case on projection stand with latch side of case toward screen. Pull on latches and separate sides of case from projector.

   One of the inside panels of the case may be used as a projection screen for small groups.

3. Set case on floor or lower shelf of stand.

4. Tie *power cord* around bottom of projection stand and connect projector to suitable power source.

5. Turn *off/fan/lamp control* to **fan** and then to **lamp.** Turn *volume control* to **on.** Adjust elevation of projector. Move projector forward or back to fill screen area with light and prefocus to get sharp edges.

   Red indicator light next to *volume control* signifies that amplifier is **on.** *Volume control* must be **on** to advance filmstrip even if an audiocassette will not be used in projector. Depress *elevation control* and raise front of projector base to desired position. Release *elevation control* to lock projection angle.

   Cooling fan must be in operation when *lamp* is **on.** On some projectors, lens is focused by rotating the *lens barrel*.

6. Remove filmstrip from storage container.

   Filmstrip should be wound so that it can be inserted into the machine head end first and with the image upside down.

   Rotate filmstrip in container by pushing against trailer or inside of strip to make it slightly smaller to facilitate removal.

7. Insert starting end of filmstrip into *filmstrip carrier*.

   Flip *film spring* forward so that filmstrip coil drops into *film cup*. Move *film speed control* to *fast*.

8. Advance filmstrip until "focus" frame appears on screen. Adjust *focus control* for sharp image.

   Depress *ADV control*.

9. Adjust *framer control* so that single frame is visible on screen.

   On some machines, the *framer control* is mounted close to the *advance control*.

10. Adjust *film speed control*.

    Control should be in **single** position.

105

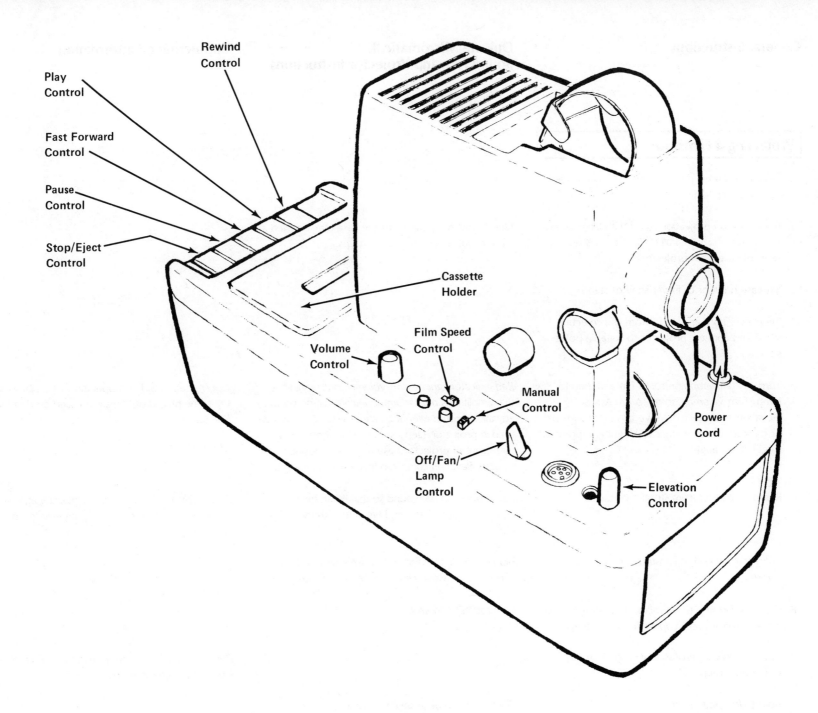

| **General Instructions** | **Dukane Micromatic II Sound Filmstrip Projector Instructions** | **Supplemental Information** |

Loading Audiocassette

11. Adjust for inaudible pulse or audible pulse playback of tape.

    Set *man/50Hz switch* to *50 HZ* if using inaudible pulse.

    If tones to signal change of visual are heard, *manual control* must be used.

12. Open *cassette holder*.

    Depress *stop/eject* key as far as it will go.

13. Insert cassette.

    Full reel of tape should be on left with exposed edge toward operator.

14. Close *cassette holder* and depress *play* key to start program.

    Volume may be adjusted with *volume control*. To stop tape during program, depress *pause* key. To resume depress *pause* key again.

15. At conclusion of program, depress *stop/eject* key only far enough to stop tape.

    Depressing *stop/eject* key to its full depth will eject tape.

16. Rewind tape.

    Depress *rewind* key.

17. Eject tape.

    Depress *stop/eject* key to full depth.

18. Rewind film.

    Move *speed control* to fast and depress *rewind* button. When completely rewound, remove film and place in protective container.

Stowing

19. Lower projector to lowest elevation.

    Depress *elevation control* and push down on front of control panel below *lens*. Release *elevation control*.

    Some machines have *thumbscrew adjustment controls*.

20. Return power controls to storage positions.

    Rotate *volume control* counterclockwise until it clicks. Move *off/fan/lamp control* to **off.** Unplug and coil *power cord* and place *cover* on projector. Secure latches.

COMMONLY USED CABLE PLUGS AND JACKS

VIDEO

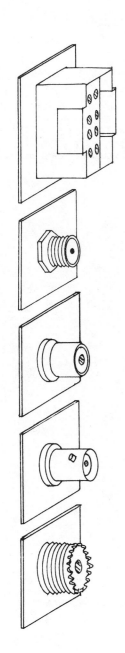

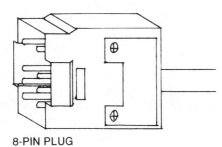
8-PIN PLUG

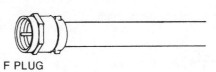
F PLUG

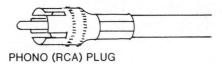
PHONO (RCA) PLUG

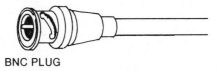
BNC PLUG

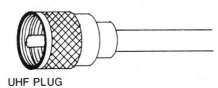
UHF PLUG

AUDIO

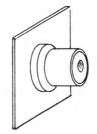

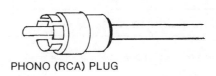
PHONO (RCA) PLUG

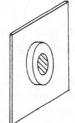

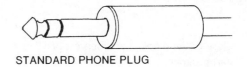
STANDARD PHONE PLUG

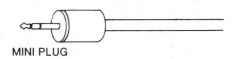
MINI PLUG

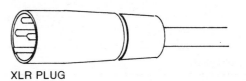
XLR PLUG

STUDIO PRODUCTION

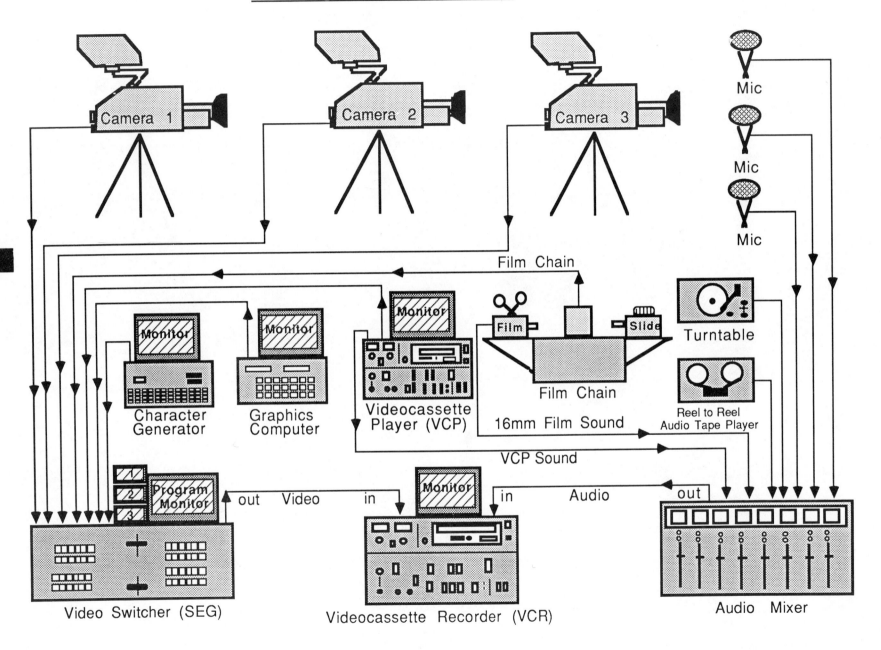

Chapter IV Video Systems

Studio Production

A television studio usually has at least two video cameras that are connected to a device called a switcher. The switcher allows the video producer to select which camera's picture is recorded onto videotape by pushing buttons located on the switcher. The producer "switches" from "camera one" to "camera two" or from "camera two" to "camera one" using this piece of equipment. Depending on the type of switcher used, the producer may use "special effects" such as dissolves or wipes in switching from camera to camera. For this reason a switcher is sometimes called a special effects generator or SEG. The illustration across the page is of a simple studio setup that displays three studio cameras connected to a switcher. On top of the switcher are three small monitors that display the output of the three studio cameras. The producer views the three monitors and makes decisions very quickly during the actual taping of the program as to which camera's picture is to be used. Once selected on the switcher, the chosen picture is simultaneously recorded and displayed on the program monitor and the VCR monitor.

There are usually more video sources connected to a switcher in addition to the studio cameras. One is a character generator. This is a computer that generates characters (letters, numbers, symbols, etc.) and allows them to be added to the program by using the switcher. The character generator has been called a video typewriter. A similar piece of hardware that offers sophisticated graphics is a graphics computer. This video source can produce very colorful graphic art, which can be stored on computer disks for later retrieval and use in other programs.

Other video sources connected to the switcher might be a videocassette player (VCP) for playing selected tape clips that are to be included in the studio program. This is often done during television news programs. A film chain usually consists of a modified 16mm motion picture projector and two 35 mm slide projectors operated by remote control from the control room of a television studio. A film chain has its own video camera, which is connected to the switcher. Film chains have a device called a multiplexer that uses mirrors to select either the film projector or the two slide projectors to project images into the video camera for inclusion in the studio program. There is normally a dissolve unit built into the film chain allowing the producer to dissolve from one slide to the next at any pace desired.

All of these various video sources meet at the switcher along with the studio cameras. The producer may decide to combine several video signals at once such as a camera showing the new Miss America being crowned, the character generator displaying her name at the bottom of the picture, and a slide inserted into the upper-right corner of the picture showing her as a little girl. This creation of video signals proceeds from the switcher to the videocassette recorder (VCR) to be recorded.

A videotape program would not be very useful to anyone without sound. The audio mixer is to sound what the switcher is to video. All audio sources meet at the audio mixer and the audio person decides, by means of control switches, which audio signal or signals proceed on to the videocassette recorder to be recorded along with the video signal. The producer uses the mixer to adjust the level of each audio source separately and to mix sound sources together. For instance, the producer could mix the output of a microphone with music from the turntable or the reel-to-reel audiotape player. Other sound sources connected to the mixer in the illustration are the 16mm film projector and the videocassette player (VCP). All of these audio sources must be monitored as carefully as the video sources to insure a high-quality program.

In summation of studio production, the chosen video and audio signals leave the switcher and audio mixer, respectively, and go to the videocassette recorder (VCR) to be recorded. This videocassette is called the master and is never distributed. Only copies or duplicates (dupes) of the master are distributed. If the VCP should malfunction and destroy part of the videotape during playback, a new duplicate can be made from the master.

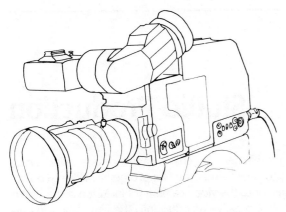
Camera and Microphone

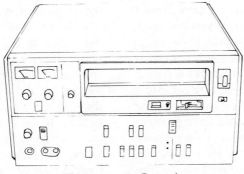
Videocassette Recorder

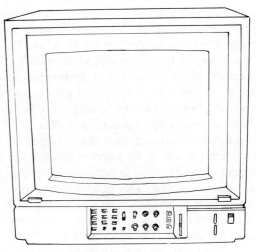
Monitor / Receiver

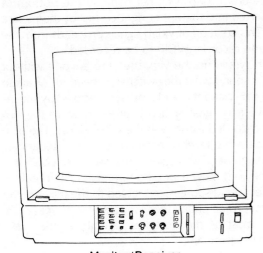
Monitor / Receiver

Input Component   Storage Component   Output Component

BASIC VIDEOCASSETTE RECORDING SYSTEM

Videocassette Recorder/Player

Video recording offers great potential for educational uses. Advantages of videotape include relatively low cost, ease of operation, erasability, immediate feedback, color, motion, and sound. Video can be used in many applications, including: (1) using a camera to magnify objects of demonstrations for a class to view on monitors, (2) viewing broadcast television, (3) recording and playing back locally produced videotapes, (4) playing commercially produced programs, (5) viewing materials at students' own pace, (6) using stop action, (7) receiving immediate feedback of recorded segments, (8) making and distributing multiple copies easily, and (9) reviewing programs. Although there are numerous formats and tape sizes available, we will concentrate on three formats, all of which are videocassettes. These three formats are 3/4", 1/2" VHS, and 1/2" Beta. Although these formats are not compatible with one another, they all offer similar controls and features. Three-quarter inch has been a popular educational/industrial format since the early 1970's. It is considered the industry standard. Half-inch systems have enjoyed widespread appeal since the late 1970's and early 1980's due to their smaller size and lower cost. A basic video recording system consists of: (1) a television camera, (2) a videocassette recorder (VCR), (3) a microphone, and (4) a television monitor/receiver.

Objectives

Observable Behavior	Conditions	Level of Acceptable Performance
1. Connect system components and make a recording.	Given a videocassette. Time limit 10 minutes.	An acceptable two-minute recording.
2. Rewind and play back recording.	Time limit 4 minutes.	Playback with properly adjusted audio and video controls on monitor.
3. Restore system to storage conformation.	Time limit 6 minutes.	All components disconnected, cables coiled, tape rewound, and returned to storage container.

Television Camera

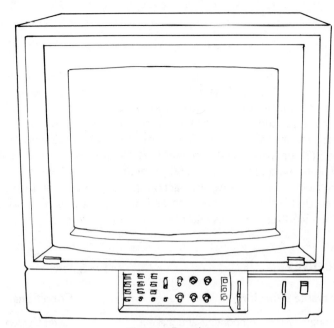
Monitor/Receiver

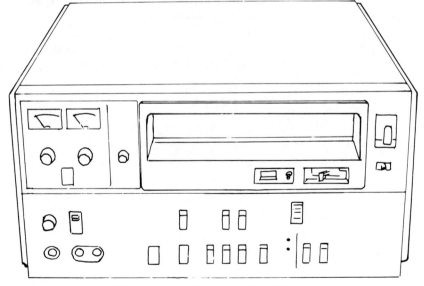
Videocassette Recorder (Front View)

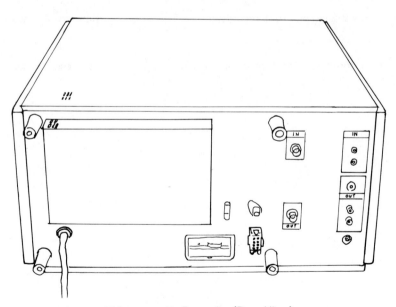
Videocassette Recorder (Rear View)

Essential Parts

Typical Video Camera

Camera head
Zoom lens
Viewfinder
Battery
AC power adaptor
Video out jack
Built-in microphone
Controls:
    Power
        Battery
        AC adaptor
    Iris automatic/manual
    Iris ring
    Focus ring
    Power zoom
    White balance
    Gain selector
    Fade time control
    Bars/camera selector
    Color temperature
        Conversion filter control

Typical Videocassette Recorder/Player

Cassette compartment
Audio level meters (Ch. 1 & Ch. 2)
Tape counter
Video in jack
Video out jack
8-pin jack
Microphone jacks
RF out jack
Headphone jacks
Audio monitor
Audio line in jacks (Ch. 1 & Ch. 2)
Audio line out jacks (Ch. 1 & Ch. 2)
Power cord
Controls:
    Power
    Stop
    Play
    Rewind
    Fast forward
    Pause
    Record
    Dub/Ch. 1
    Search (FWD & REV.)
    Tracking
    Skew
    Audio limiter
    Audio monitor
    Input select
    Eject

Typical Monitor/Receiver

Video line in jack
Audio line in jack
Video line out jack
Audio line out jack
8-pin jack
VHF input jack
UHF input jack
Power cord
Controls:
    Power
    Picture
    Color
    Hue
    Bright
    Volume
    Channel number
    Input select (TV-VTR-LINE)
    75-ohm termination

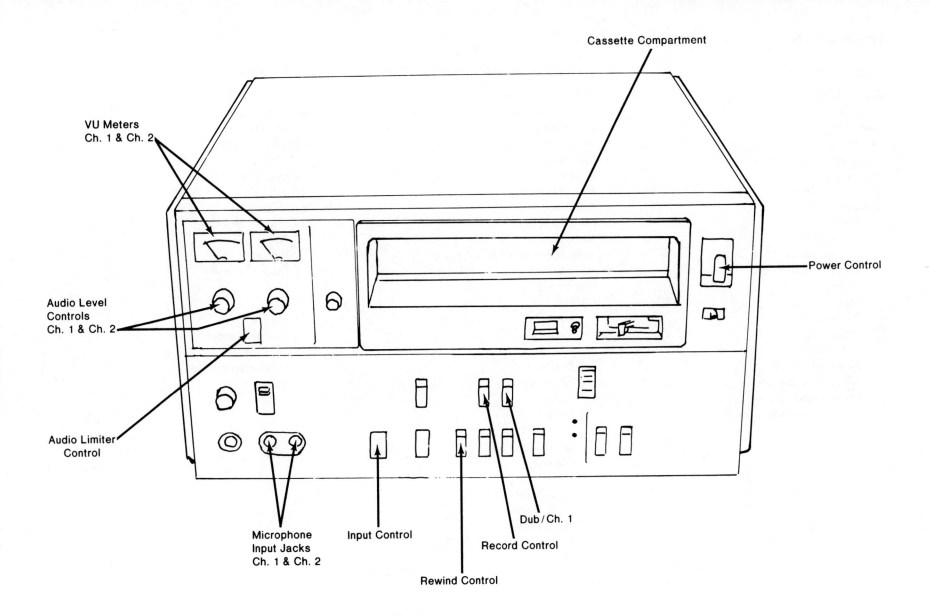

| **General Instructions** | **Sony VO-5600 Videocassette Tape Recorder Instructions** | **Supplementary Information** |

Setting Up Recorder

1. Read General Equipment Operation Principles, Audio Principles, Projection Principles, and AV Equipment Principles before proceeding.

2. Tie *power cord* around bottom of cart or table leg and connect to a suitable power source.

3. Turn power on.

   Depress *power control* to turn on the VCR.

4. Load cassette into recorder.

   Position videocassette with label facing up and tapered corners toward machine and push videocassette forward as far as it will go into the *cassette compartment*.

5. Rewind videocassette, if necessary.

   Depress *rewind control*. When videocassette is wound, machine will stop automatically.

   Always rewind a videocassette before use. The previous user might have neglected to completely rewind it.

6. Connect microphone.

   On Sony VO-5600, connect microphone to *microphone input jack* on front panel of machine. Use the one marked Ch. 2.

   Some machines have a *microphone input jack* on rear panel. Most videocassette recorders have two audio recording tracks. Individual or stereo tracks may be recorded or erased after original video recording has been made. *Ch. 1* can be used for audio dubbing.

7. Adjust *audio level control*.

   Depress *record control* and adjust *Ch. 2 audio level control* so that needle on *VU meter* does not peak beyond red block on meter while talent is rehearsing. *Audio limiter control* may then be used by turning on.

   The *audio limiter control* prevents distortion that might be caused by unusually loud audio input.

8. Set *input control* to line.

   The only time the *input control* would be set on TV would be to record an off-air broadcast from a television set. All other uses, such as live recording or dubbing, will be on line input.

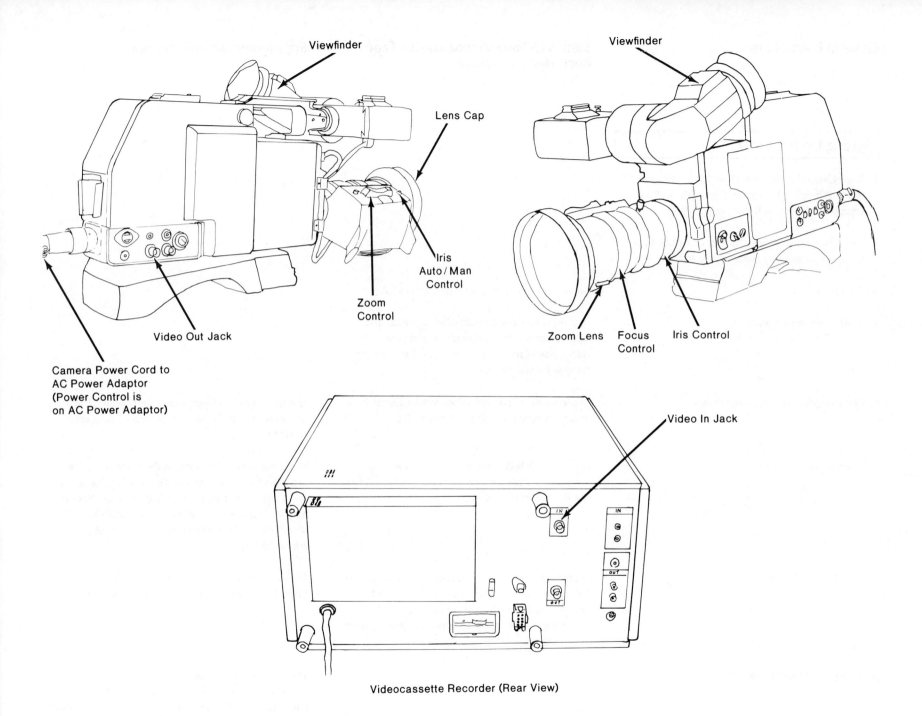

| **General Instructions** | **Sony DXC-1820 Color Video Camera Instructions** | **Supplementary Information** |

Setting Up Camera

9. Connect video cable from camera to recorder.

                                                                             Connect video cable from *video out jack* on camera to *video in jack* on recorder.

10. Connect camera *power cord* to suitable power source, usually the AC power adaptor.

                      Turn on the Sony CMA-7 power adaptor.   Several seconds after activating *power control*, picture tube in *viewfinder* will reveal a light gray glow.

11. Remove *lens cap* from front of *zoom lens*.

                                                                             TV camera *lenses* should always be capped when not in use to prevent damage to pickup tube. This is true whether *power control* is activated or not. Avoid aiming camera at extremely bright light sources.

12. Frame scene and check for any lighting problems.

                    Set *iris auto/manual control* on the *zoom lens* to auto. The iris will automatically adjust to the brightness of the scene. Normally the auto position is used.   For manual adjustments, set the *iris auto/manual control* to manual and turn the *iris control*. This may be beneficial when recording a person against a light or dark background.

13. Focus camera.

                    Focus is obtained by moving *zoom control* to telephoto (maximum close-up) and adjusting *focus control* until subject appears sharply defined on *viewfinder*. Adjust *zoom control* to wide angle (maximum picture coverage) to ensure that all the scene can be included. If necessary, adjust coverage by moving camera.   If *zoom lenses* are focused at extreme telephoto (close-up), critical focus will be maintained when control is moved to extreme wide angle.

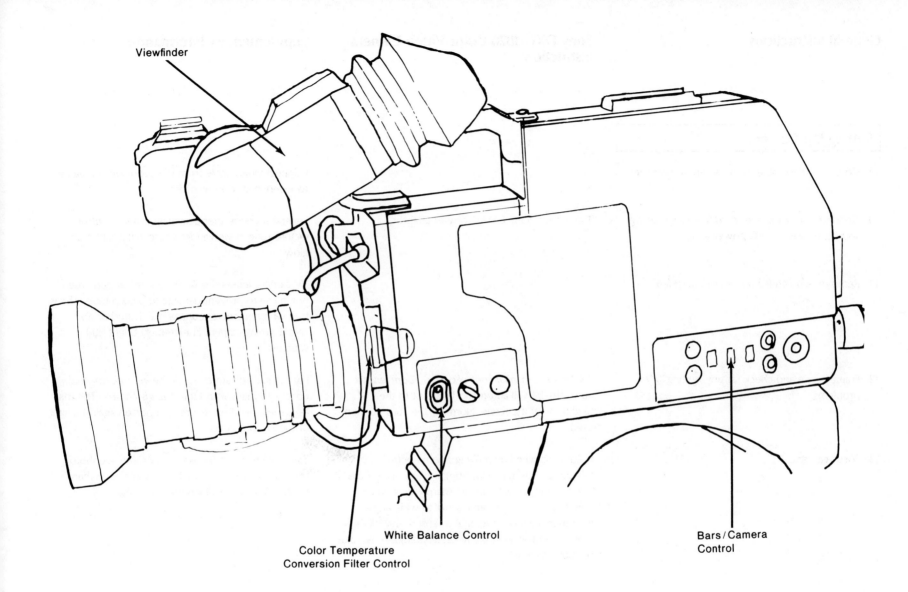

General Instructions

14. White balance the camera for correct color reproduction.

Sony DXC-1820 Color Video Camera Instructions

Set the *color temperature conversion filter control* to the proper position determined by the lighting conditions. Switch *bars/camera control* to camera. Zoom in to a white card reflecting the same light as that which will be used during recording, and set the *white balance control* to auto. The white balance will automatically adjust illuminating the white balance indicator in the *viewfinder*. The *white balance control* automatically returns to its center position when released.

Supplementary Information

The color temperature of light varies with the time of day, the weather conditions outside, and the type of lighting inside. Video cameras use internal filters to compensate for different lighting conditions. Some of these conditions may be: sunrise, sunset, bright sun, cloudy, rainy, fluorescent light, and Iodine light.

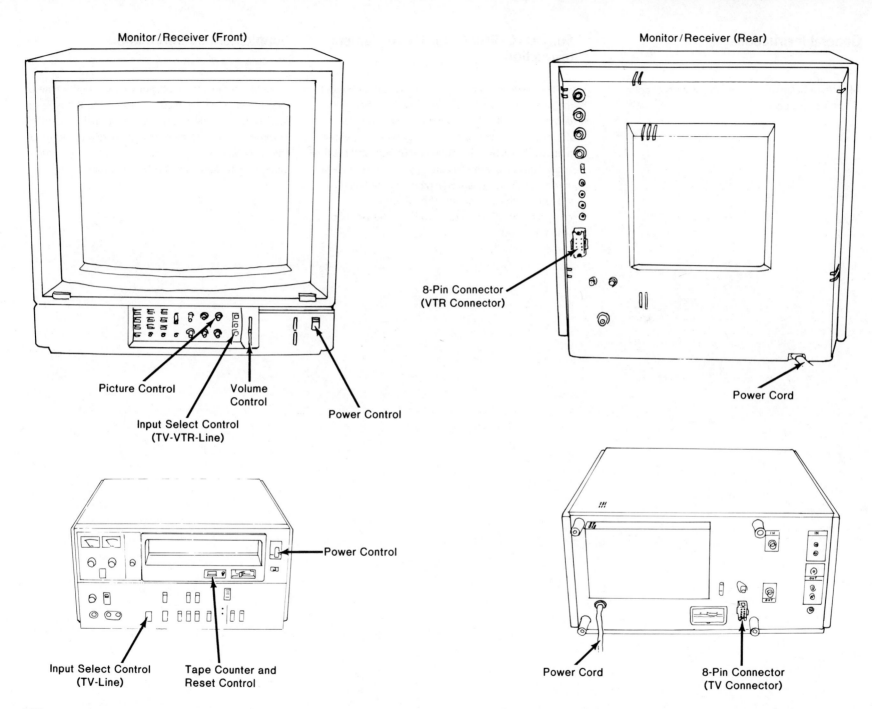

| General Instructions | Sony CVM-1270 Color Monitor/Receiver Instructions | Supplementary Information |

Setting Up Monitor

15. Connect videocassette recorder to *monitor/receiver*.

    This is accomplished by using a single 8-pin cable to connect *8-pin connector* on videocassette recorder to *8-pin connector* on monitor.

    Some models of recorders and monitors do not have an *8-pin connector* so separate video and audio connections must be made.

    NOTE: It is possible to record without connecting monitor; however, it is necessary if you wish to check quality of recording by playing back a brief taped sequence before proceeding.

16. Connect *power cord* from monitor to suitable outlet.

    Failure to set *input select control* will result in no picture on monitor. Most TV receivers designed for educational use have an *input select control* so they may also be used as off-the-air receivers when set on TV.

17. Turn monitor *power control* on.

18. Set *input select control* on monitor to VTR.

19. Set *tape counter* to 000.

    Push *reset* button.

    The numbers indicate number of revolutions of take-up reel. These numbers are not consistent from one VCR to another.

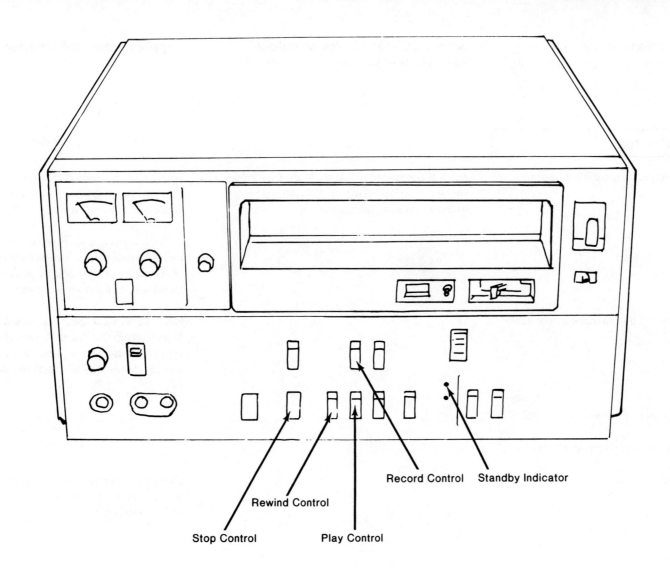

General Instructions	**Sony VO-5600 Videocassette Tape Recorder Instructions**	**Supplementary Information**
20. Start recording.	Simultaneously depress *record* and *play controls* to initiate recording.	A little thought will enable the operator to determine which controls are to be activated. Makes and models will vary in operational procedure. On some machines, a *standby light* is not included; on these types it is not possible to depress another command function until the VCR has completed its present cycle. On some models the VCR will internally delay the new command until the current function is completed.
21. Signal talent to begin speaking into microphone while you operate camera.		Extraneous sounds will be recorded; so directions to talent or camera operator should be silent.
22. Stop recording when finished.	Depress *stop control*.	
23. Rewind videocassette to starting point.	Wait for *standby light* to go out; then depress *rewind control*.	Never disconnect power cord while tape is moving.

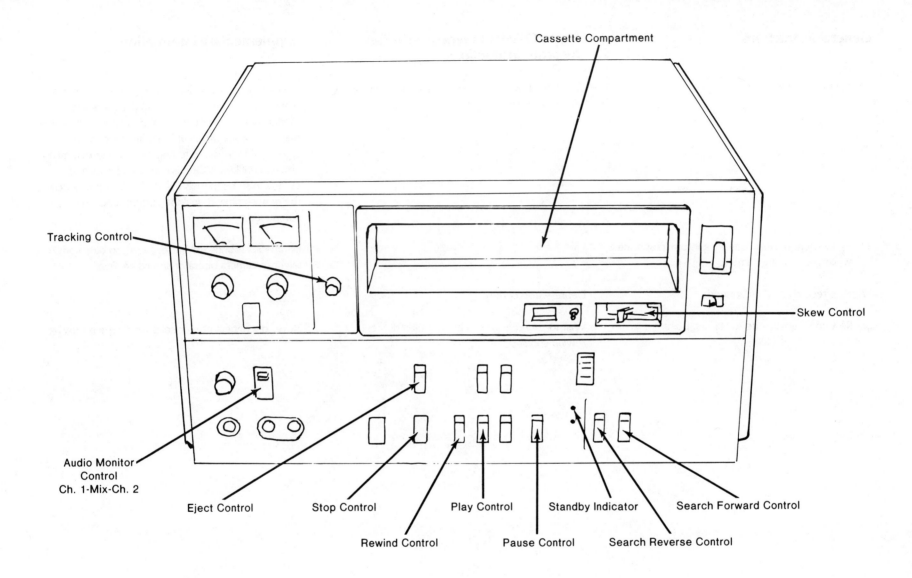

| **General Instructions** | **Sony VO-5600 Videocassette Tape Recorder Instructions** | **Supplementary Information** |

Playing a Videocassette

24. Place videocassette in recorder.	Insert videocassette into the *cassette compartment* and push it forward until the machine accepts it.	NOTE: Camera need not be connected during videocassette playback.
25. Select appropriate audio channel.	Move *audio monitor control* to Ch. 2 or mix.	
26. Make sure videocassette is completely rewound.	Depress *rewind control*. When videocassette is rewound, machine will stop automatically.	
27. Play videocassette.	Depress *play control*.	To view the picture at five times normal speed either in forward or reverse, depress the *search controls*. The *pause control* may be used to temporarily stop the videocassette without unthreading it. On some recorders the *pause control* will enable the viewer to freeze an image on the monitor.
28. Adjust *tracking control*, if necessary.	Horizontal image break-up indicates tracking maladjustment. Rotate *tracking control* until picture on monitor is clear.	
29. Adjust *skew control*, if necessary.	If a flagging or hooking distortion appears in the upper portion of the picture, move the *skew control* slowly to the left or right to correct.	Videocassette played back on same machine that it was recorded on probably will not require adjustment of *tracking* or *skew controls*.
30. Adjust picture and sound controls on monitor.	Adjust *picture* and *volume controls*, if necessary.	The *picture control* adjusts picture brightness, contrast, and color at the same time.
31. At conclusion of tape, stop VCR.	Depress *stop control*. Wait for *standby indicator* to go out.	
32. Rewind tape.	Depress *rewind control*. When videocassette is rewound, machine will stop automatically.	
33. Remove tape from VCR.	Depress *eject control*. Videocassette will eject from machine.	

Television Camera

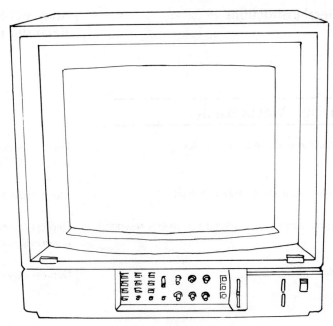
Monitor / Receiver

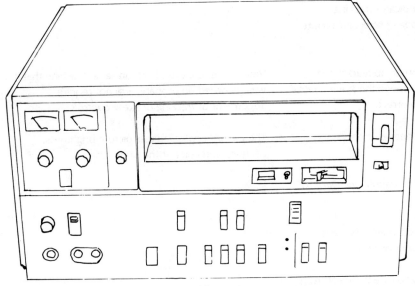
Videocassette Recorder (Front View)

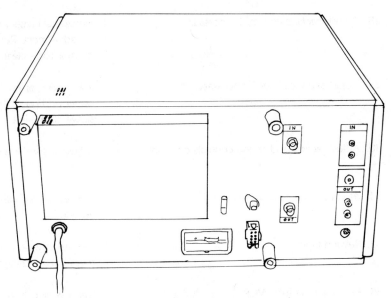
Videocassette Recorder (Rear View)

General Instructions

Sony VO-5600 Videocassette Tape Recorder Instructions

Supplementary Information

Stowing

34. Restore all components to storage conformation.

Replace *lens cap* and turn all power controls off. Disconnect and coil all cables.

A TV camera lens should be capped when not in use regardless of whether the power is turned on or not.

ELECTRONIC FIELD PRODUCTION

Shooting

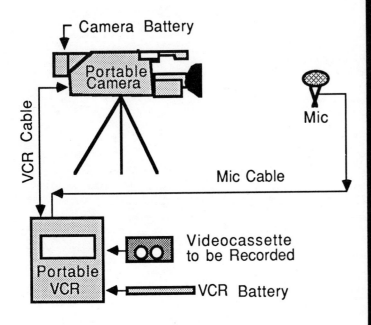

Note: This could be a 3/4 inch, 1/2 inch VHS or 1/2 inch Beta portable videocassette recorder.

Editing

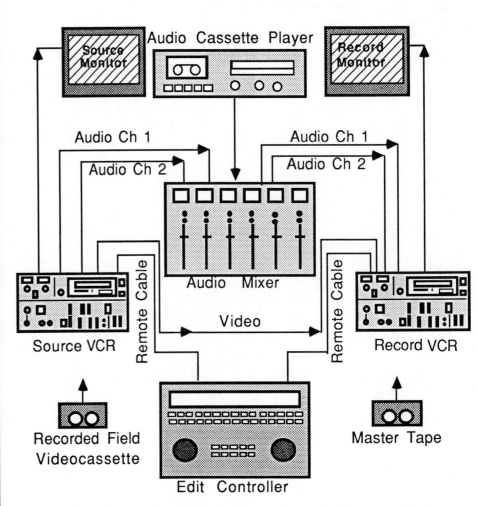

Note: This could be a 3/4 inch, 1/2 inch VHS or 1/2 inch Beta editing system.

Electronic Field Production

Read Studio Production before proceeding.

The term electronic field production pertains to portable, battery-powered video cameras and videotape recorders that can be taken out into the field to produce videotape programs outside of the studio "on location." This video equipment is of the same design that revolutionized television news operations during the 1970s when the CBS Television Network began replacing its newsfilm cameras with small battery-operated television cameras and videotape recorders. They found that this equipment allowed them to gather the news for broadcast faster and more economically than with film. At the same time, they could go anywhere with it that they could go with a film camera. This technological revolution was labeled electronic newsgathering or ENG.

The use of this ENG equipment by producers outside of television news became known as electronic field production or EFP. This is a very different production style compared with traditional studio production. With EFP, a program is shot "film-style" in the field with one camera and videotape recorder and then edited together later on a videotape editing system. Because of its similarity to filmmaking, EFP is sometimes called electronic cinematography. In studio production, by means of the switcher, the producer edits the program as it is being videotaped (production editing). In EFP, the producer edits after the shooting is completed (post-production editing). Clearly, post-production editing offers the producer more time for editorial decision making compared with the often pressure-filled environment of production editing in a studio with three cameras and a switcher.

The illustration at left displays a portable, battery-powered video camera and videocassette recorder with a microphone for shooting in the field. The videocassette editing system consists of a source VCR with a monitor, a record VCR with a monitor, an edit controller, an audio mixer, and an audiocassette player. The source and record machines are operated remotely using the edit controller. The field videocassettes are brought in and played back on the source VCR. The program is pieced together, one scene at a time, onto a blank videocassette (master tape) in the record VCR. The scenes on the field tapes can be rearranged on the master tape in any sequence the producer desires. There are two channels of audio available on the master tape — channel 1 and channel 2. Narration, music, sound effects, etc., may be edited into the program from either the source VCR or the audiocassette player. The "insert editing" feature of the record VCR allows the producer to insert video onto the master tape and later insert audio (narration perhaps) onto channel 1 and different audio (music) onto channel 2 without harming the video. This capability provides endless creative possibilities.

The Camcorder has become a popular option to the separate camera and recorder in the 1/2 inch VHS and Beta formats. A Camcorder consists of a video camera and a videocassette recorder in a single, battery-powered unit. Because of their compactness and convenience, they have surfaced in both the consumer and educational marketplaces. These units use the same videocassette tape as previous VHS and Beta VCRs. A VHS or Beta Camcorder could easily be inserted into our illustration in place of the portable camera and VCR under "Shooting." However, the quality of VHS and Beta would not be as high as 3/4 inch U-matic.

New videotape formats are being developed continually. The VHS manufacturers now offer Super VHS (S-VHS) with VCRs that produce over 400 lines of resolution compared with the 240 lines of conventional VHS and the 260 lines of standard 3/4 inch U-matic. S-VHS uses a new videotape oxide formulation to help achieve this greater sharpness and improved color. A S-VHS machine will play back a normal VHS videocassette at the lower resolution but a S-VHS recording will not play back on a normal VHS machine. The higher resolution and lower price compared with the current 3/4 inch U-matic makes S-VHS an attractive alternative in the future video marketplace.

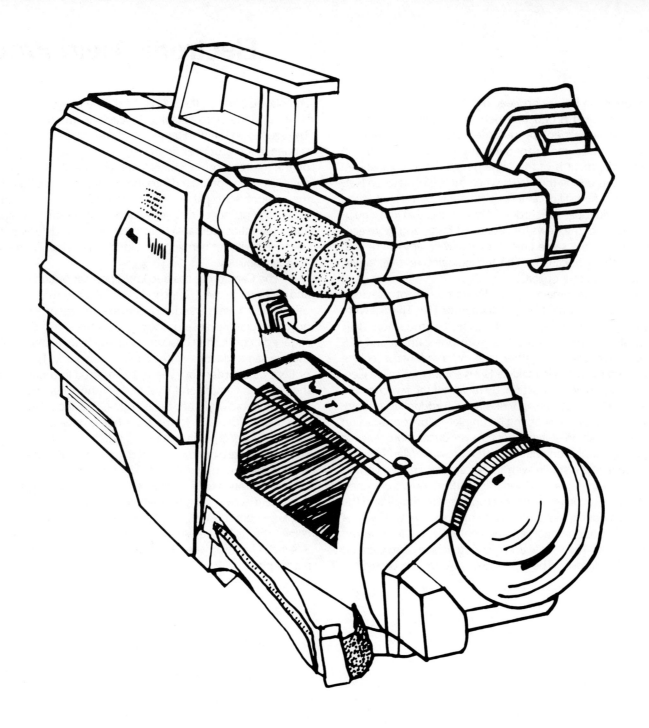

The ½" VHS Camcorder Recorder/Player

Video camcorder units offer ease of use and lightweight portability. The versatile ½" video camcorder lends itself to educational and business applications. A camcorder may be used in business to record business proceedings, meetings, and training sessions and in education to record lectures, demonstrations, classroom activities, field trips, and counseling sessions.

A tape recording made with the camcorder may be played back immediately after rewind. It may be edited or stored for later use. The camcorder becomes a valuable teaching tool when it is used to record lectures, counseling sessions, demonstrations, and field trips for later reference or assignment for individuals or groups of learners. The following step-by-step procedures will demonstrate how to set up and record with a camcorder. The ½" VHS recording units have gained widespread popularity since the late 1970's because of their light weight and ease of operation. A basic VHS camcorder system consists of: (1) camera and recorder, (2) battery pack, (3) AC adaptor/battery charger, (4) connecting cables, and (5) storage case.

Objectives

Observable Behavior	Conditions	Level of Acceptable Performance
1. Set up and operate the camcorder unit to make a satisfactory recording.	Given a VHS videocassette. Time limit 10 minutes.	An acceptable two-minute recording.
2. Rewind and play back recording.	Time limit 3 minutes.	Playback tape for evaluation on camera viewfinder/monitor.
3. Disassemble camcorder system and accessories and return to storage.	Time limit 10 minutes.	All components disconnected and returned to storage case.

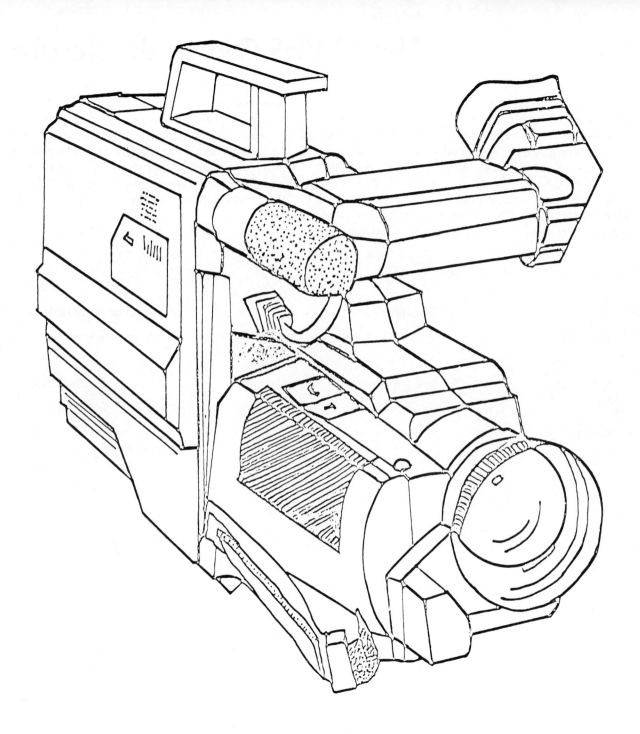

Essential Parts

Typical Camcorder

AC adaptor
AC adaptor connector
Battery pack
Battery compartment
Cassette compartment
Carrying handle
Electronic viewfinder
Hand grip
Lens
Lens cap
Power on indicator
Recording indicator lamp

Controls:
    Auto focus mode selector
    Battery pack eject
    Data function selector
    Eject
    Eyepiece corrector
    Fast forward
    Focus mode selector
    Focus ring
    General operation
    Macro close-up
    Manual focus ring
    Manual zoom
    Pause/still
    Power with indicator
    Power zoom
    Record
    Record start/stop
    Rewind/search
    Standby function
    Start/stop
    Stop
    Tracking
    TV/VTR
    White balance
    Zoom
    Zoom ring lever

Additional on AG-160 Panasonic Camcorder

Fade in/fade out
Backlight
High-speed shutter selector
Tape running
Battery eject
Accessory shoe
Built-in microphone
Earphone jack
Standby indicator
VCR control cover
White balance sensor window

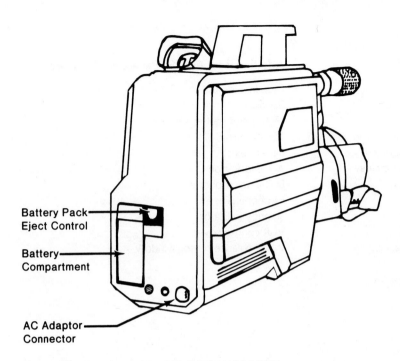
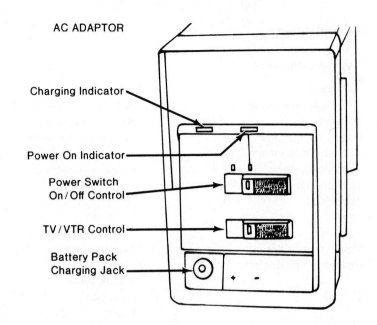
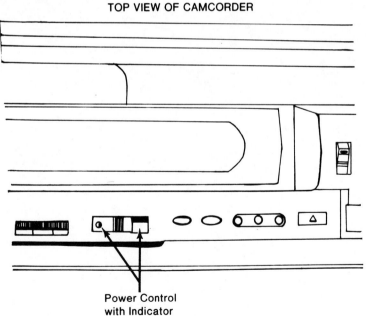
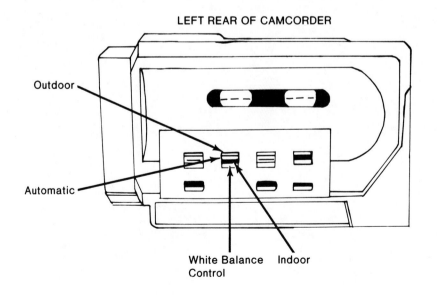

| **General Instructions** | **AG-160 Panasonic Camcorder Instructions** | **Supplementary Information** |

Setting Up

1. Read General Equipment Operation Principles, Audio Principles, Projection Principles, and AV Equipment Principles before proceeding.

2. Unpack camcorder and AC adaptor or battery pack.

3. Select power supply.

    To use AC power:
    a. Plug AC adaptor connector into AC adaptor connector jack on rear of camcorder.
    b. Attach AC cord to power outlet.
    c. Turn power switch control on.
    d. Glow of charging indicator will signal unit is on.

    Set *TV/VTR control* to VTR.

    For battery power:
    a. Insert fully charged battery into *battery compartment* until it clicks into place.
    b. Arrow located on top of battery pack should be up and pointing into *battery compartment*.

    Use battery power when operator wishes to be free from power cords. A fully charged battery should power unit up to two hours. Use camcorder's standby feature to save battery power. Recharge battery after use.

4. Turn power on.

    Slide *power control* to rear of camcorder. Power-on indicator will glow.

5. Set *white balance control*.

    Set *white balance control* to indoor, outdoor, or auto.

    Either indoor or outdoor settings should be used to allow sensors to function correctly. Auto should be considered only for transition scenes from indoor to outdoor or outdoor to indoor. Auto setting may not be desirable under some conditions:
    a. Either camcorder or subject is in shade.
    b. Facing extremely bright sources.

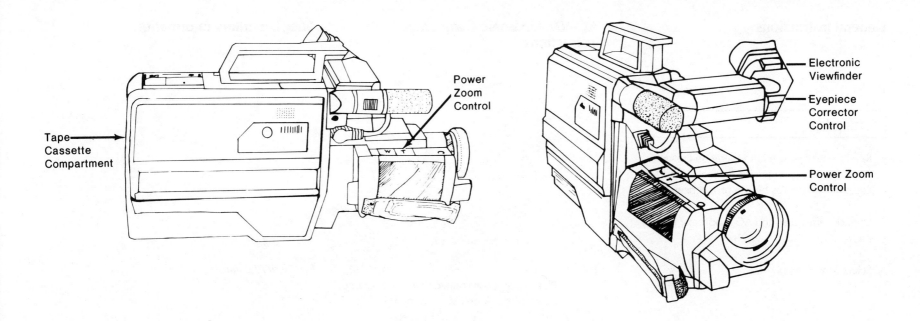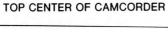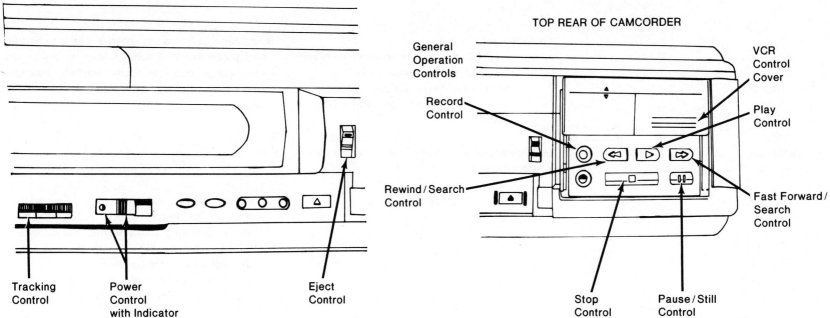

General Instructions	**AG-160 Panasonic Camcorder Instructions**	**Supplementary Information**
		c. Sunrise or sunset.
		d. Light source outside sensing range of camcorder such as snow scenes, candlelight, etc.
6. Remove *lens cap*.		
7. Open *tape cassette compartment*.	Depress *eject control*.	
8. Insert tape cassette.	Tape should be inserted with arrow pointing down with window side out. Tape erasure prevention tab should be in place, otherwise record mode will be canceled.	½" VHS tape of varying quality is available. High-quality tape is an absolute must because maintenance and repair costs for electronic equipment will cancel apparent savings on tape purchases.
9. Close *tape cassette compartment*.		
10. Locate *general operation controls*.	*Tracking, power,* and *eject controls* will be needed eventually.	*Tracking control* allows precise adjustment of tape path in machine to accept tape recorded on another camcorder. If you plan to use a tape made on another camcorder, always play enough tape to set *tracking control* for best picture with least horizontal interference top and/or bottom.
11. Locate *tape running controls*.	*Record, rewind/search, play, fast forward/search, stop,* and *pause/still controls* will be used first. *Power zoom control* should also be located. Slide *VCR control cover* up.	
12. Rewind tape.	Depress *rewind/search control*.	Tape should always be rewound to make certain that recording is started at beginning of tape unless there is continuation of a tape.
13. Extend *electronic viewfinder*.	Rotate *electronic viewfinder* and turn outward until it locks with a click. *Eyepiece corrector control* allows correction to suit operator.	

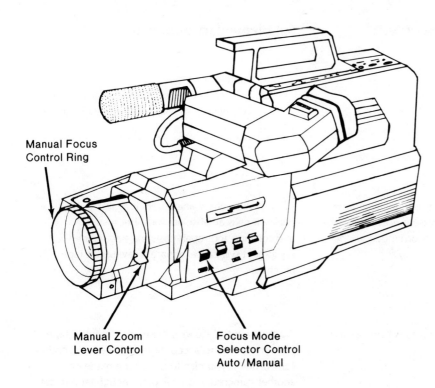
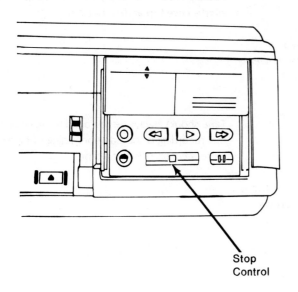
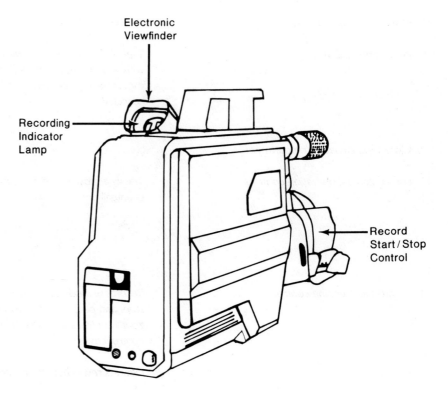

| **General Instructions** | **AG-160 Panasonic Camcorder Instructions** | **Supplementary Information** |

14. Set *focus mode selector control*.

Auto focus may be used if subject fills most of picture area. It may be adjusted to read smaller, more select areas of the picture by depressing *auto control* repeatedly to gain full, 2/3, or 1/3 frame reading.

In manual focus mode, momentary automatic focus adjustment is possible for as long as *auto control* is depressed. Rotate *manual focus control ring* to manually adjust focus. Operator may choose automatic or manual zoom.

Under most conditions, camcorder lens should be focused while at telephoto (close-up) position. Subject will then remain in focus as lens is moved toward wide-angle position.

15. Macro close-up capability.

Set the *focus mode selector control* to manual. Rotate *manual zoom lever control* to click stop. When control is turned farther, lens is in macro range. Turn *zoom lever control* slowly and focus on subject.

Recording

16. Start recording.

Slide *camera/VTR control cover* down over VCR controls. Allow a few seconds for camcorder to go into pause mode.

Depress *start/stop control*. The *recording indicator lamp* in the *electronic viewfinder* will light up as recording starts.

A red light on left side of *viewfinder* will be on in addition to the orange light.

17. Stop recording.

Depress *start/stop control*; *recording indicator lamp* will go out as recording stops and camcorder goes into recording pause mode.

If camcorder is left in pause mode longer than 5 minutes, it will automatically switch to standby mode (viewfinder is off) to protect tape and conserve battery power. Depress *standby control* to go back to recording pause mode.

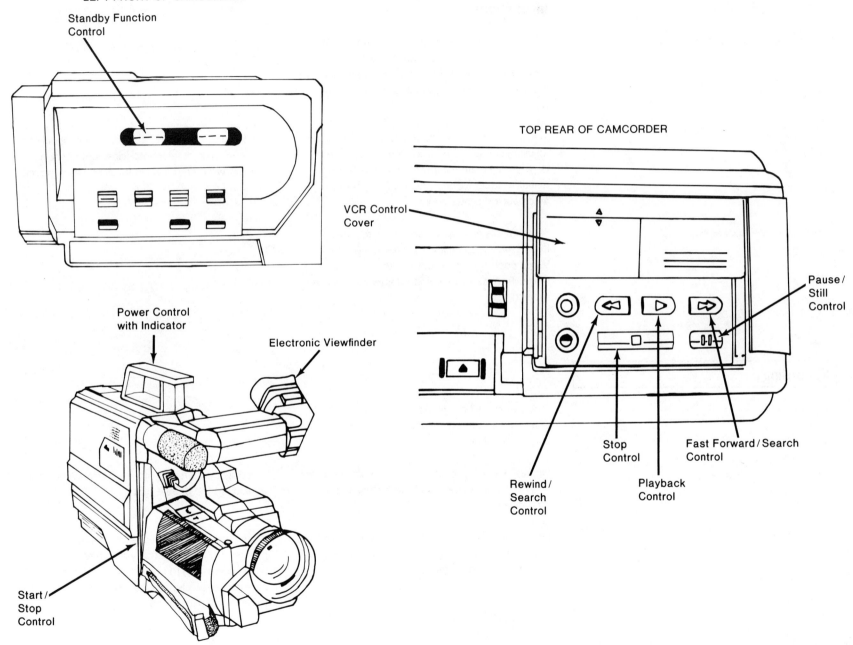

142

General Instructions	AG-160 Panasonic Camcorder Instructions	Supplementary Information
18. *Activate standby function control.*	Depress *standby control* and corresponding lamp will light. (No picture on *electronic viewfinder*) Wait at least 4 seconds before performing any other operations. Depress *standby control* again to prepare camcorder for recording next scene.	It takes about 5 seconds for image in *electronic viewfinder* to become stable after taking unit out of standby mode by activating *standby function control*.
19. Make a test recording with talent.		

Playing Back Through Electronic Viewfinder

20. Stop tape.	Depress *start/stop control* once.	
21. Rewind tape.	Slide up *VCR control cover* on top of camcorder. Tape will unthread and stop. Depress *rewind control*. Tape will rewind and stop at beginning.	Make certain that *edit control* is off.
22. Play back.	Depress *play control*.	When *fast forward/search control* or *rewind/search control* is depressed while camcorder is in playback mode, tape will be played back in forward or reverse direction at three times normal speed.
23. Still playback.	Depress *pause/still control* once.	To continue normal playback, depress *pause/still control* again.
24. Stop playback.	Depress *stop control*.	
25. Rewind tape.	Depress *rewind/search control*. Tape will stop at beginning of recording.	

Stowing

26. Disassemble and restore to storage conformation.	Turn *power control* off, place *electronic viewfinder* in storage position, and return all accessories as well as camcorder to carrying case.	

CHAPTER V   Troubleshooting Guide for Instructional Equipment

This should be helpful in diagnosing and correcting difficulties encountered in operating instructional equipment. Problems have been listed under the following categories:

1. General
2. Visual System
3. Audio System
4. Transport System
5. Videotape System

	PROBLEM	POSSIBLE CAUSE	REMEDY	EXPLANATION or SUGGESTION
General	Nothing happens when machine is turned on.	No power—not plugged in.	Check power plug for snug fit. Try reversing plug.	Power plug may have accidently been pulled. Worn or damaged outlets or plugs are often at fault.
		No power in the building circuit.	Try another electrical appliance in the same outlet.	Safety fuse or circuit breaker may have deactivated the circuit; usually a result of overload or electrical short circuit.
		No power in the machine circuit.	Push the red circuit breaker button.	Some pieces of equipment have a circuit breaker on the control panel that breaks the circuit if an overload or an electrical short circuit is apt to damage the machine.
		Faulty cord or wiring in machine.	Try another power cord from a similar machine.	If it still does not operate, the problem is in the machine wiring. Borrow replacement machine and notify person in charge.
		Safety device on machine.	Close or tighten access cover. Check threading pattern to determine whether clutches or automatic shut-off mechanisms have been tripped.	Safety interlock is designed to prevent operation when cover is not properly secured or machine is not correctly threaded.

		PROBLEM	POSSIBLE CAUSE	REMEDY	EXPLANATION or SUGGESTION
Visual System	Lamps	Motor or fan runs but lamp does not light.	Lamp switch **off**. Bulb improperly installed. Bulb may be burned out. Faulty switch.	Activate switch. Replace lamp.	Jiggle switch to see if lamp flickers. Gently move switch to different position to make contact. If lamp lights momentarily, switch is defective. Borrow replacement machine and notify person in charge.
	Focus	Screen image is fuzzy and not clear.	Lens not focused. Lamp incorrectly installed.	Adjust focus.	On some projectors, internal lenses or bulbs may be improperly aligned. Notify competent technician.
	Brightness		Dirty lens. Film not properly positioned in film channel.	Clean lens with soft cloth or lens tissue. Recheck threading.	If film is not properly aligned in channel of 16mm projectors, new sprocket holes may be punched.
	Others	Inverted image on the screen.	Slide upside down. Filmstrip improperly threaded.	Invert slide in carrier. Rethread filmstrip.	Lenses invert images. The first frames of the filmstrip must be inserted in the carrier upside down.
		Edge of film visible on screen.	Motion picture film not properly positioned in film gate.	Stop projector and rethread.	If the gate is closed with film only partially within the gate area, there may be damage to the film.
		Multiple, blurred, or unstable images.	See transport system problems.		
		Parts of two pictures or frames are seen on screen.	Framer adjustment.	Adjust framing control.	

Bulb Changing Suggestions

When changing a burned out bulb in an overhead projector, be careful not to touch glass part of bulb because oil and moisture from skin may damage replacement bulb. To replace flange-base or bayonet-base bulbs, press down on old bulb while turning it counterclockwise. Spring pressure will usually cause it to move upward slightly so that it may be grasped for removal. Align the replacement bulb with slots in socket, press down, and turn clockwise. To remove pin-base bulbs, press down on lever located in projector lamp housing. If there is no lever, grasp bulb and pull upward. Replacement bulb should be lined up with key slot and pushed into place.

		PROBLEM	POSSIBLE CAUSE	REMEDY	EXPLANATION or SUGGESTION
Audio System	Poor Fidelity	Low-pitched, "gutteral."	Machine set at wrong speed.	Adjust speed control.	Sound recorded at a faster speed than playback.
		High-pitched, "chipmunk-like."	Machine set at wrong speed.	Adjust speed control.	Sound recorded at a slower speed than playback.
		Audiotape is mushy and indistinct.	Dirty playback head.	Clean head.	A cotton swab with alcohol will clean off the deposit of metallic oxide. If problem still exists, the recording could be of poor quality.
		Kodak Pageant 16mm sound is mushy and indistinct.	Soundtrack not focused.	Fidelity control lever improperly adjusted. Try center position while projector is running.	Exciter lamp beam must be focused on soundtrack for best sound reproduction.
			Film not snug around the sound drum.	Adjust threading conformation.	

	PROBLEM	POSSIBLE CAUSE	REMEDY	EXPLANATION or SUGGESTION
Audio System — No Sound or Low Volume	Low volume.	Audiotape or film not threaded properly.	Check threading.	Film may not be moving through components of sound system. Tape may not be in contact with playback head.
		Wrong side of tape might be in contact with playback head.	Rewind and rethread tape.	
		Dirty playback head.	Clean head.	A cotton swab with ethyl alcohol will clean off deposit of metallic oxide.
		Microphone used for recording was not compatible with the recorder.	Use the proper microphone.	Microphones furnished with recorders are of matched impedance; high or low. Substitute components must match.
	No sound being produced.	Speaker cord not connected.	Plug speaker in.	
		Amplifier not on.	Switch amplifier on.	Amplifier switch is frequently separate from the master switch.
		Faulty microphone, amplifier, or speaker.	Try another microphone or speaker and cord from a similar machine.	Borrow replacement machine and notify proper authorities.
		Exciter lamp on 16mm sound projector burned out.	Replace lamp.	

		PROBLEM	POSSIBLE CAUSE	REMEDY	EXPLANATION or SUGGESTION
Transport System	Audio	Tempo is slightly off, resulting in distortion of pitch, rate, etc.	Dirty or faulty drive mechanism.	Have the machine cleaned.	Belts, gears, etc., need occasional cleaning or replacing.
		(See Audio System Problems Section)	Some object rubbing on record or tape reel.	Remove resistance.	
		Record repeats same phrase.	Faulty record.	Move tone arm toward center of record.	Damaged grooves.
	Projectors	Carousel projector tray cannot be removed.	Tray identification number not on 0.	Turn power control on. Remove tray manually.	Turn tray removal screw with a coin. It is located in center of top of projector. Lift tray off while turning screw. Turn tray over (if lock ring is in place) and adjust bottom plate to allow retainer to engage bottom plate.
		Chattering, clicking noise emitting from projector; picture may be jumping on screen.	Lost film loop.	Depress loop forming roller to reset film loops to conform to threading path on projector.	Restores lower loop while projector is in operation. Torn or worn sprocket holes in the film can be at fault.
			Film gate open.	Close film gate.	Pressure plate must be pressed against film in the film gate.
			Sprocket clamps open.	Close drive sprocket clamps.	Clamps prevent film from "crawling" during operation.
		Film piles up on the floor.	Film not winding on take-up reel.	Secure to take-up reel.	End of the film leader should be inserted into the slot in the take-up reel; several turns of film leader will secure it in place.
			Take-up reel does not turn or turns too slowly.	Avoid contact of take-up reel with table. Check drive belts for correct positioning.	Take-up reel must not contact a table or cabinet. Reel must turn freely.

	PROBLEM	POSSIBLE CAUSE	REMEDY	EXPLANATION or SUGGESTION
Videotape System	No picture on the monitor/TV.	Lack of good connection from playback deck.	Try another cable.	Either an 8-pin or coaxial video cable must carry signal to monitor/TV.
		The last videocassette tape user did not rewind tape.	Rewind tape.	
		Monitor/TV set may have multifunction selector switch or maybe set to an incorrect position.	Set selector to correct position.	Most monitor/TV sets have three-position-selector switches for VTR, LINE, and TV. For an 8-pin cable, use VTR.
		Record button on video recorder not activated.	Push record button to see image on monitor/TV set.	Recording does not start until forward or play button is simultaneously activated with record.
		Brightness control on monitor/TV set turned to black or off.	Adjust control.	
		Tape will not move through recorder or playback deck.	Notify proper authorities of need for technical repair.	
	Videotape stops and VCR controls fail to operate.	Lack of leader tape on the take-up reel.	Eject videocassette and take up leader tape manually by turning the take-up reel in the direction of the arrow on the bottom of the cassette shell. Then reinsert the videocassette.	

	PROBLEM	POSSIBLE CAUSE	REMEDY	EXPLANATION or SUGGESTION
Videotape System	Recorder will not record new material.	Record button not activated.	Always push record and forward or play buttons at same time.	You may think that both controls are being pushed simultaneously when actually the record function can be canceled by pushing forward or play first.
		Accidental erase prevention button removed.	Replace red safety button.	With red accidental erase prevention button removed, recorder will not record new material.
		Lack of good connection from camera to recorder.	Try another cable.	
		Cable from camera to recorder improperly connected.	Should be connected to video input jack on VCR.	
		Blank videotape.	Try a new tape and notify videotape dealer.	Defective videocassette.
		Input select control not properly set.	Set selector to correct position.	Input select switch must be set to **line** to record signal from video in jack on VCR.
	A videocassette cannot be inserted.	Another videocassette has already been inserted.	Eject old videocassette.	
	Sound cannot be heard from a monitor or headphones.	Audio monitor control in wrong position.	Set the audio monitor control to the proper position.	This control can be set to either Ch. 1, mix, or Ch. 2.

	PROBLEM	POSSIBLE CAUSE	REMEDY	EXPLANATION or SUGGESTION
Videotape System	Power does not turn on.	The camcorder is in the standby mode.	Press the standby button.	The standby mode conserves battery power.
	Power turns off as soon as it is turned on.	Condensation is present in the camcorder.	Wait until moisture has evaporated.	Condensation usually forms when the camcorder is brought into a warm room on a cold day.
		Battery is discharged.	Replace with fully charged battery.	Charge the battery for at least 3 hours.
	Power is on and no picture appears in viewfinder.	Camcorder is in standby mode.	Press the standby button.	
		Lens is capped.	Remove the lens cap.	
	Color is poor.	Improper white balance setting.	Adjust white balance setting.	
		Poor lighting.	Use adequate light and light of the correct color temperature.	

	PROBLEM	POSSIBLE CAUSE	REMEDY	EXPLANATION or SUGGESTION
Videotape System	Snow in video.	Clogged video heads.	Clean U-matic heads with a foam swab and generous quantities of cleaning fluid. Do not spray fluid directly at heads or drum. For 1/2-in VHS VCRs, use a nonabrasive cleaning tape.	Particles of metallic oxide from tape will rub off and lodge in the tiny gap of the electromagnets that are the video heads. Sometimes caused by leaving VCR in pause more than two minutes.
	Video, when played back, has a disruption in the image at evenly spaced intervals.	Electrical interference.	Remove power cord and reinsert in wall outlet the opposite way.	Sometimes the electrical polarity is reversed and the only remedy is to change the direction of current flow from wall outlet to VTR.
		VTR out of adjustment. Drive motor probably does not maintain a constant speed.	Notify proper authorities of need for technical repair.	Tape speed through a video recorder must maintain a constant speed during recording and must be played back at exactly the same speed.
	Tape does not move through videotape recorder.	VTR may be colder than room temperature.	Always allow VTR to warm gradually to room temperature. Often it will take 3 to 4 hours.	A VTR that is colder than room temperature will collect moisture and cause videotape to adhere to guides and the drum on which video heads make contact with tape.

	PROBLEM	POSSIBLE CAUSE	REMEDY	EXPLANATION or SUGGESTION
Videotape System	Camcorder will not record.	The videocassette tab is broken out.	Cover tab hole with adhesive tape.	The videocassette tab is used to prevent accidental exposure.
	Fuzzy picture.	Subject is out of focus.	Zoom in to ECU and focus camcorder.	The manual focus mode is recommended.
	Camcorder does not play back on TV set.	Camcorder is not connected to TV set correctly.	Connect camcorder to the TV set correctly using a coaxial cable from the TV jack on the AC adaptor to VHF "in" on the rear of the TV set.	Make sure the TV/VTR selector on the AC adaptor is set to VTR.
		TV set is not on correct channel.	Adjust TV set to the proper playback channel.	The Panasonic AG-160 uses channel 3 or 4.
	Playback picture is distorted.	Tracking is out of adjustment.	Adjust tracking control until picture clears up.	This sometimes occurs when playing back a videotape that was recorded on another VCR.

Chapter VI MICROCOMPUTERS AS AUDIOVISUAL TOOLS

Chapter Objectives

After reading this chapter, the student will be able to:

1. Describe the essential parts of a microcomputer.
2. Describe how microcomputers can be used as projection devices.
3. Describe how microcomputers can be used to prepare classroom documents.
4. Describe the instructional applications of microcomputers.

Needs

As an audiovisual tool, a microcomputer can meet the following needs:

1. The need to project computer-generated displays to an audience.
2. The need to easily produce and revise documents.
3. The need to provide interactive or one-on-one instruction to students.

Overview

Computers are defined as information processors. Computers have the ability to receive information (input), process the input according to internal instructions, and finally display the results as output.

Like cassette tapes, overhead projection systems, and filmstrips, computers can be used to transmit information to an audience. Usually computer information is transmitted or displayed on monitors much like television screens or through information printed on paper as printouts or hard copies.

A Multipurpose Machine

Most pieces of audiovisual equipment, such as audiocassette tape recorders, are single-purpose systems; however, computers may be multipurpose systems that can do many different tasks. The audiocassette tape recorder can only record and play back sound. A multipurpose computer can at one time be a sophisticated graphics tool to create drawings and charts while at another time the computer can be a personal tutor teaching a preschooler the alphabet. What allows the computer to assume different roles and abilities are the various programs that are used to control the computer system. Programs, or software, are the detailed instructions that control the entire computer system.

Typical Components of a Microcomputer System

Software

The term *microcomputer* describes a small personal computer system that can usually fit on top of a desk. Most microcomputer software is stored on a diskette and usually comes with a manual called documentation, which contains instructions for the computer user on using the program. Computer software normally comes on 5.25 inch or 3.5 inch diskettes. Information is recorded magnetically on diskettes much like sound or video is recorded on cassette or videocassette tape. To insure against any damage occurring to the floppy disk, do not touch any of the exposed magnetic area, keep the disk away from objects that generate magnetic fields such as telephone receivers or small motorized appliances, and read any other precautions that come with the diskette.

It is important to remember that software controls every aspect of the computer's operations; software should be given great consideration when selecting or evaluating a computer system.

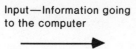

Input—Information going to the computer →

Computer

The Central Processing Unit controlled by a program process input

Output—Processed information coming from the computer →

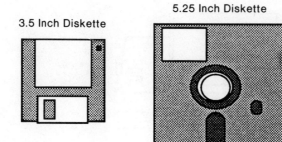

3.5 Inch Diskette

5.25 Inch Diskette

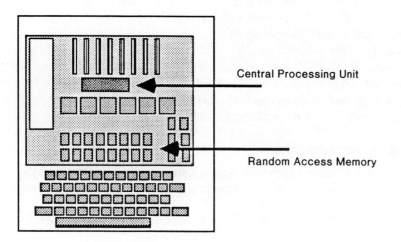

Central Processing Unit

Random Access Memory

Hardware

Hardware describes the physical components of the computer system. Following are typical pieces of hardware found in most computer systems.

Central Processing Unit

The most important component in a computer is the Central Processing Unit (CPU). The CPU performs all of the computer's internal calculations and decision-making routines as instructed by the software.

Random Access Memory

Random Access Memory (RAM) works closely with the CPU. The computer's CPU uses RAM as a storage area to temporarily store information going to or coming from the CPU.

Computer memory is measured by units called *bytes*. One byte is equal to one character. For example, the word *computer* would occupy eight bytes in the computer's memory. When measuring the computer's memory often the letter K, for Kilo, is borrowed from the metric system to represent approximately 1000 bytes. A computer that is described as having 64 K or 64 K bytes of RAM has the capacity of storing approximately 64000 characters in Random Access Memory. Diskette capacity is also measured by bytes. Diskette capacity normally varies from 120 K bytes to 800 K bytes depending on the system that is used. Storage capabilities on newer, more powerful computer systems are being measured by Mega bytes, millions of characters and by Giga bytes, billions of characters.

Information in RAM is temporary and volatile; information is lost once the computer is turned off. It is important to save your work often to a permanent, nonvolatile medium such as a diskette.

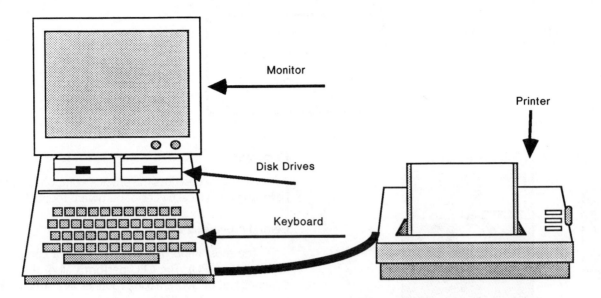

Keyboard

The keyboard, which looks much like a typewriter keyboard, is an input device that is used to send information to the CPU.

Disk Drives

Disk drives are used to read from or write to diskettes. A disk drive is both an input and output device that has the capability of sending information from the diskette to the CPU or sending processed output from the CPU or memory to be saved on a diskette.

Monitor

The monitor is an output device that displays information on a screen like a television set. Monitor capabilities vary greatly from monochrome monitors, which only produce alphanumeric text in one color, to color-graphics monitors, which can display a variety of colors and can also display graphics such as drawings, pictures, and charts on screen.

Printers

Printers are output devices that the computer uses to print information onto paper. The quality and capabilities of printers vary from printer to printer. Low-cost dot-matrix printers produce characters through combining little points or dots. When higher quality printing is needed, more-expensive, letter-quality printers such as daisy wheel or laser printers can produce full-formed characters.

```
This is an example of dot-matrix printing.
If you look close, you can see that characters
are formed by combining little points or dots.
```

```
This is an example of letter-quality
printing.  Characters are full formed.
```

Supplementary Information

When selecting a computer system, care must be taken to match the hardware with the software. Some questions that should be addressed are:

1. Does the software require color-graphics capabilities? If yes, does the computer system have color-graphics capabilities?
2. What is the minimum amount of memory the software requires to operate? Does the computer system have the required amount? Can the memory of the computer be increased to meet future needs?
3. Does the software require any extra attachments or peripherals such as a printer? Does the software support the printer that is going to be used?

Using a Microcomputer as a Projection Device

A computer can be used to project information to an audience much like an overhead projector or motion picture projector. Large-screen monitors are commonly used in the classroom to allow the audience to view computer output. Projection devices and panels that connect the computer to an overhead projector are common ways that computer output can be displayed on a wall or portable screen.

Presentation software allows the creation of computerized slide shows and presentations that use color and/or animation.

Typical Components

A microcomputer system
A projection device
Presentation software

Some advantages of using a computer as a projection device with presentation software are:

1. The presenter can demonstrate computer programs to an audience.
2. Once an image is created, it can easily be changed. Modifications such as changes in lettering size, correcting spelling errors, changing the position of text, changing the font or type style can be accomplished without having to reenter the entire text.
3. Computers can be very accurate in positioning text, margins, and graphics.
4. Once an image is produced it can be incorporated into other images.
5. Depending on the presentation software used, special effects such as fade-ins, color, and animation can be incorporated in the presentation.

Using a Microcomputer to Prepare Classroom Materials

A microcomputer can improve the quality of documents that are used in the classroom. The typical hardware and software used to create printed originals for classroom handouts are:

A microcomputer system
A printer
A word processing and/or graphics program

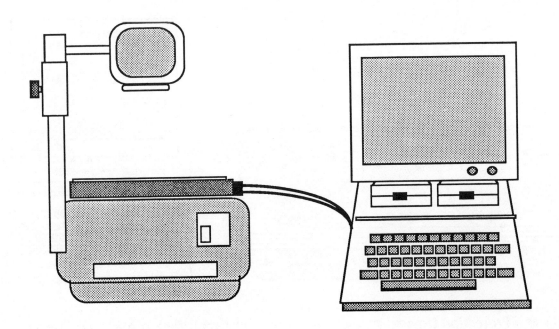

IBM Corp

Word processors are programs that allow the computer to be used like a typewriter to produce documents such as letters and classroom handouts. A computer used with word processing software can improve the process of creating and revising documents by allowing the user to easily manipulate text, margins, and other aspects of the printed page. Once a document is created and saved to a diskette, it can always be loaded back into the computer to correct any mistakes, add or delete information, or adjust margins. Some word processing systems allow for using various fonts or type styles and different letter sizes within the same document.

Graphics programs are used to create and manipulate drawings and shapes. Graphics programs allow for the computer user to accurately create and position boxes, circles, lines, and other objects on the printed page. Like a word processing document, once a graphics picture is created and saved to disk, it can be loaded back into the computer for changes or revision.

Printed output from both word processors and graphics programs can be reproduced as thermal spirit masters or transparencies.

Some advantages in using a computer to prepare printed presentations are:

1. Once a document or picture is created, it can be revised and reprinted.
2. Various type styles (fonts) and character sizes can be generated.
3. Computers can be very accurate in positioning text, margins, and shapes.

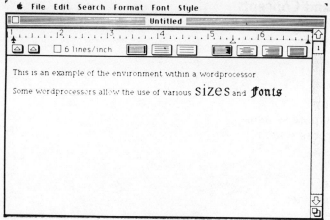

1. MacWrite

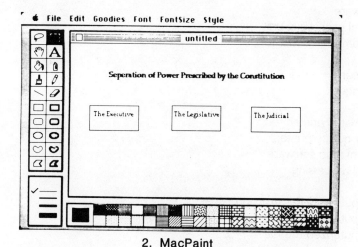

2. MacPaint

Using a Microcomputer as an Instructional Tool

Many educators are using computers as supplements to teaching to provide interactive instruction to individuals or small groups. Computers have the ability to display information to a student and also receive and evaluate information provided by the student. This area of computing is referred to as *computer-aided instruction* (CAI). There are many types of instructional software that can be divided into four general categories.

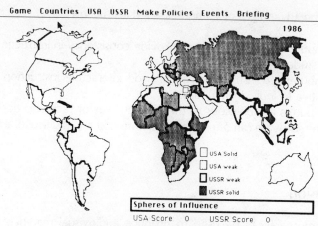

Copyright © 1985 Chris Crawford
All Rights Reserved

Computerized Drills

Computer drills are used to reinforce skills already taught. The student is presented with practice problems by the computer and the student responds accordingly.

Computerized Tutorials

Computerized tutorials are used to teach new skills or concepts to the student. The computer is like a private tutor presenting new information to the student.

Computerized Simulations

Computerized simulations allow the student to interact with real-life situations to learn or reinforce a concept or skill. Mimicking dangerous experiments, learning how to fly an airplane, and reproducing historical decision-making processes are common uses of computerized simulation.

Computerized Instructional Games

Computerized instructional games instruct through presenting the student with fun tasks. Instructional games usually incorporate fantasy situations and present rewards or points when appropriate behavior is displayed by the student user.

Some advantages of computer-aided instruction are:

1. Computerized instruction can provide consistency in instruction from one student to the next.
2. Computerized instruction can provide interactive instruction that is sensitive to individual abilities.
3. Computers are oblivious to the students' sex or race and depending on the software can offer consistent nonsexist and multicultural instruction.

Conclusion

Computers can be used to produce audiovisual materials for the classroom teacher. Although there are many advantages in using a computer for generating audiovisuals, a computer is a complex, expensive machine and appropriate time should be spent on learning the particular computer system that will be used. The advantages of using a microcomputer should be weighed against the time and effort that goes into learning about and operating a computer system. It is important to remember not to use a computer for the sake of using a computer but to use a computer because it can meet a unique need that only the computer can satisfactorily fulfill. Do not use a computer to project information to an audience when a simple overhead transparency or other less complex system will meet the particular need.

Terms and Concepts

Computer
Input
Output
Monitor
Hard copy
Single-purpose system
Multipurpose system
Graphics
Programs
Software
Microcomputer
Diskette
Documentation
Hardware
Central processing unit
Random access memory
Byte
Keyboard
Disk drive
Printer
Dot-matrix printer
Letter-quality printer
Presentation software
Word processing
Font
Computer-aided instruction
Computerized drill
Computerized tutorial
Computerized simulation
Computerized instructional game

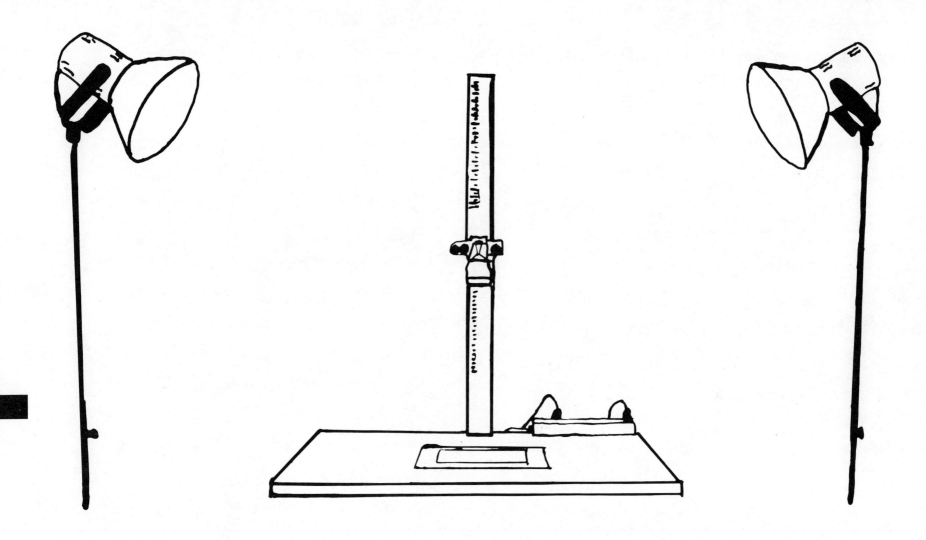

Chapter VII Photographic Copystand Project

A copystand and camera combination is useful when it is necessary to produce slides or negatives from opaque printed or originally designed materials. The copystand is essential for preparing animation sequences or titles for motion pictures. Effective use of a copystand and camera requires an elementary knowledge of lighting, lenses, and apertures and the relationship of aperture, shutter speed, and focus. The purpose of this assignment is to give the student certain elementary information and the opportunity to manipulate it.

Objectives

Observable Behavior	Conditions	Level of Acceptable Performance
1. Set up lights, camera, and materials to be photographed.	In lab.	Lights properly placed and camera in appropriate relationship to material to be photographed.
2. Photograph your name, I.D. number, and lab section. This original may be created with stencils, dry-transfer lettering, felt-tip pen, or India ink and pen.	In lab.	Must be in sharp focus, have correct exposure, and be framed tightly around the image.
3. Photograph two other visuals of your choice.	In lab.	Must be in sharp focus, have correct exposure, and be framed tightly around the image.
4. Restore copystand to storage and return camera with film to Media Circulation counter.		Turn off lights, remove camera from stand, and remove all personal belongings from stand.

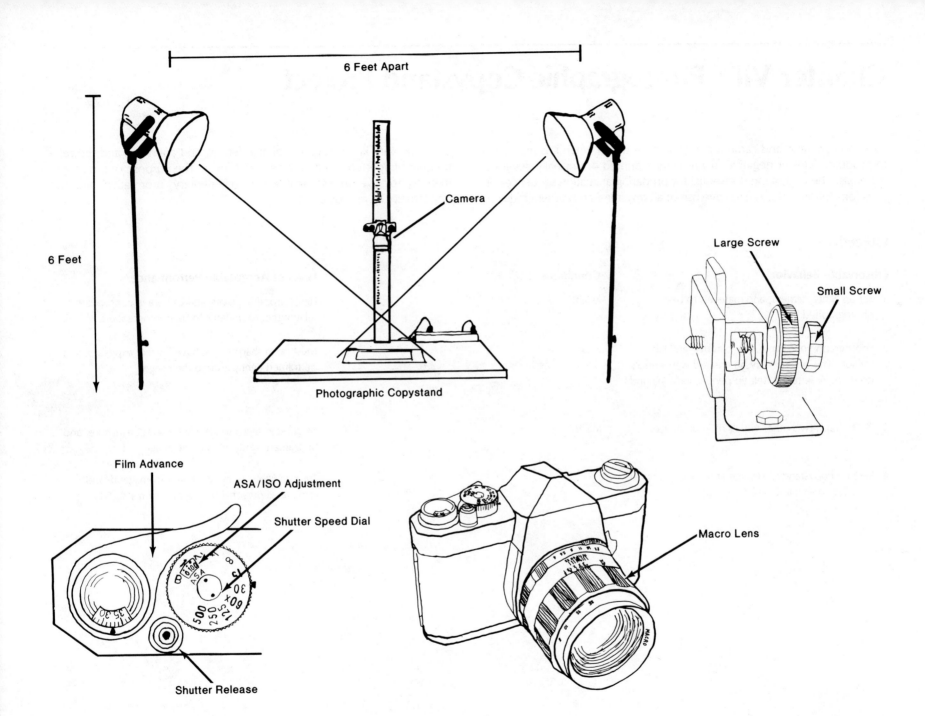

General Instructions

1. Connect copystand lamps to power outlets.

2. Adjust lamps.

3. Attach camera with *macro lens* to copystand.

4. Set *ASA adjustment* for film being used.

5. Turn on lamps and turn off room lights.

Pentax Instructions

Turn *small screw* into camera base until it stops. Tighten with *large screw*.

Pull *shutter speed dial* toward you, rotate, and release it at correct setting.

Supplementary Information

Some models are equipped with adjustable lamp arms.

Lamps should be 6 feet in height, 6 feet apart, and at a 45 degree angle to material on copystand. The "hot spot" of each lamp should fall just beyond the material to be photographed so that it is lighted by the "spill" from both lamps. Failure to do this will result in uneven lighting.

Copystand configurations may vary, but it is important to have the camera firmly attached so that it does not move when making *aperture* and *shutter adjustments* or advancing film.

Any camera that has an automatic metering system will have some means of adjusting to a variety of films.

ASA/ISO is a numerical system of rating the sensitivity of film to light, or how quickly film accepts light; the film speed.

By turning off room lights, reflections from shiny surfaces are minimized.

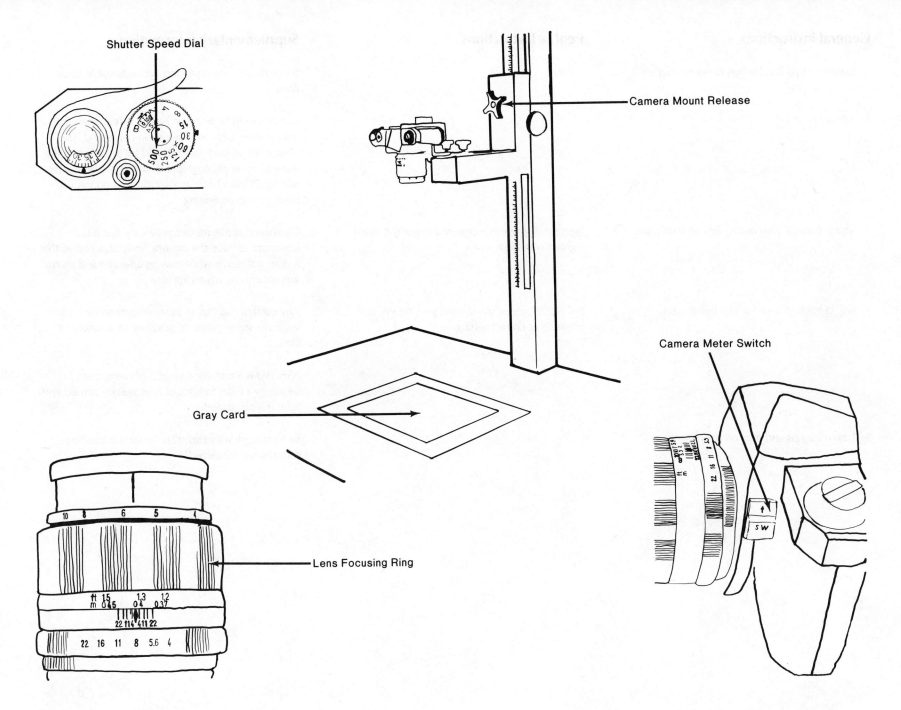

| **General Instructions** | **Pentax Instructions** | **Supplementary Information** |

6. Adjust camera height so that material is properly framed.

Camera will record slightly more than what is seen in viewfinder, so image must be cropped on all sides.

Hold on to camera with left hand and use the right hand to turn release on copystand counterclockwise to unlock camera mount. It is spring-loaded, so hang on to camera mount and do not let it hit you in the face.

7. Frame and focus by adjusting lens or adjusting camera distance to materials.

Rotate *lens focusing ring* to sharpen image. If framing is incorrect, readjust height and focus controls.

Focus again if you change camera distance to materials.

8. Set shutter speed by turning *shutter speed dial* to a number.

Shutter settings must be made only on the numbered intervals.

Shutter speed dial controls the length of *time* the shutter is open. Shutter is a metal or cloth shield in camera that protects film from light. When shutter is opened, film is exposed.

9. Turn on *camera meter switch*.

Located on right-hand side of camera body and near the lens.

Some camera models have a meter switch activated by a stand-off position of film advance lever, shutter release button, or lever located on camera body.

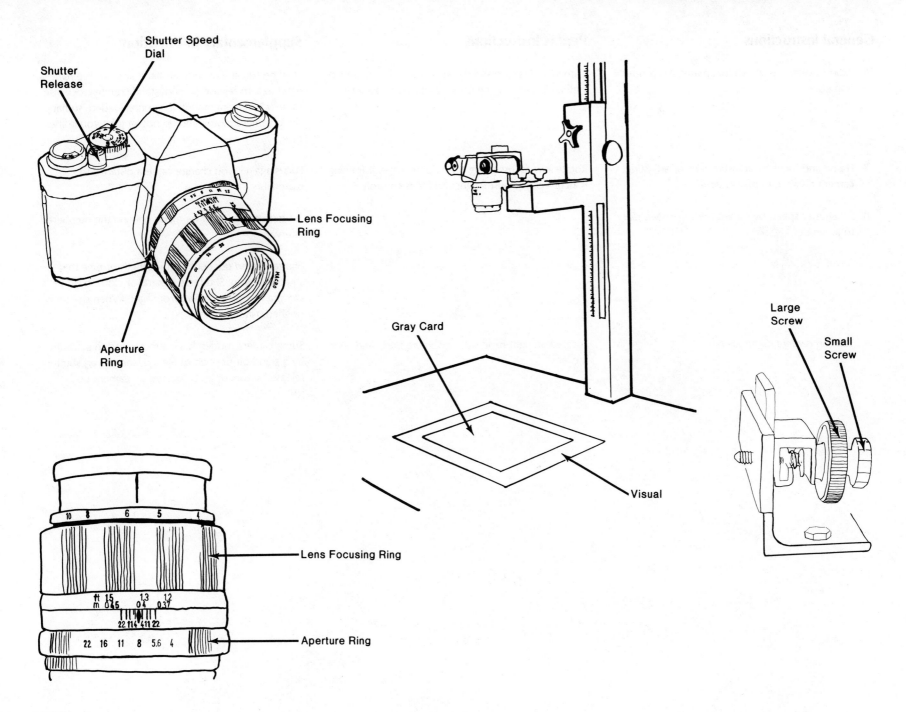

General Instructions	**Pentax Instructions**	**Supplementary Information**
10. Use *gray card* to determine correct exposure. A *gray card* is used to reflect an *average* of the amount of light falling on a visual rather than the light that reflects off specific light or dark areas.		Aperture (lens opening) is a small hole in the lens through which light travels to reach the film. Size of aperture is either fixed or adjustable depending on camera. The f-stop numbers on the *aperture ring* indicate how large the aperture opening is; the larger the f-stop number, the smaller the lens opening.
11. Look through camera viewfinder and adjust *aperture ring* on lens until correct exposure is indicated.	Needle should fall in middle of notch on left side of viewfinder for correct exposure. *Shutter speed dial* may need to be adjusted. Turn but do not pull it toward you.	Adjustment of shutter speed may be necessary to achieve a correct exposure setting.
12. Remove *gray card*. Recheck framing and focus of material before squeezing *shutter release*.		When removing *gray card*, visual may move slightly.
13. Squeeze *shutter release* and advance film for next exposure.		
14. Repeat steps 6 through 13 for each photograph.		
15. Restore copystand to storage and return camera to Media Circulation counter.	Rotate *large screw* to unlock camera body and then back *small screw* out of camera base plate.	Hang on to camera as it is being removed from stand. Take your time.

Chapter VIII Lettering Principles

The effectiveness of visual materials and displays is to a large degree dependent on the quality of the lettering in titles and labels. Legible and attractive lettering can be produced by almost anyone who is willing to invest a small amount of time and effort in practice. To guide the student, the following fundamentals are offered.

Visibility of lettering is determined by size and contrast. Legibility is controlled primarily by style and spacing.

Size

Letters for nonprojected display materials should be at least 1/4" high (lowercase x) when viewed at a distance of eight feet. Uppercase letters should be larger. They should be proportionately larger if they are to be viewed from greater distances. Lettering on transparencies for use on the overhead projector should be at least 1/4" high (lowercase x).

Contrast

The contrast between letters and their background is an important factor in their visibility. Light-colored letters on a dark background or dark letters on a light-colored background are most visible. Do not confuse light with bright, nor dark with dull.

Style

The style used determines how easily the letters can be read. Letters that are simple are more easily read than complex styles. Generally avoid the use of serifs.

𝕵𝕴𝕸 JIM
TOO COMPLEX   SERIFS

For titles, short statements, and labels, it is advisable to use UPPERCASE (CAPITAL) LETTERS. When you are using long sentences (more than six words) or paragraphs, lowercase letters combined with capitals are more legible than capitals alone.

JIM   jim
UPPERCASE   LOWERCASE

One of the most common errors in lettering is failure to keep the tops and bottoms of letters even, particularly when lettering freehand. Lightly drawn guidelines will help keep lettering uniform in height and can be erased after inking.

Always use at least two GUIDELINES

Spacing

Spacing refers to the amount of space left between the letters of a word, between the words themselves, and between the lines. A neat lettering job can be ruined by improper spacing. There are two types of spacing:

MECHANICAL SPACING involves treating the separate letters as if they were in a box or retangular block. The spacing is determined by the equalization of the *distances* between the blocks. This type of spacing, while relatively easy to do, is generally less legible and attractive.

OPTICAL SPACING should be used when feasible, because the result is much more legible and attractive. It is accomplished by equalization of the *spaces* between letters. The alphabet can be divided into three types of letters: (1) regular or rectangular letters such as B, E, H, I, M, N, R, S, and U, which are easy to space. (2) The circular letters C, D, G, O, and Q (two O's together show more white space at the top and bottom than at their closest points), which must be placed closer to their neighbors to equalize the white space between the letters, and (3) the irregular letters such as A, F, J, K, L, P, T, W, X, Y, and V, which are difficult to space.

Centering a line of lettering is difficult for the novice. Two suggestions are offered: (1) if you have only a single line to be centered, work in the center of a sheet that is larger than required, then trim to desired size; (2) using tracing paper, make a quick rough of the desired letters to determine the amount of space required.

In addition to spacing letters with care, you should avoid crowding between words and between lines. There are general rules for spacing lettering, but the best way to learn is through practice. Spacing is simply the arrangement of letters and words so that they look attractive to the eye.

Dry Transfer

Professional-looking lettering for instructional materials can be produced with dry-transfer letters, known also as "press-on," "rub-on," and "transfer type." The letters are printed on a plastic carrier sheet. Letters are available in black, white, and a variety of colors. Some of the colored letters are transparent and will project in color if adhered to a piece of acetate and used as a transparency on the overhead projector. Some of the letters have a wax adhesive and some have a heat-resistant adhesive. The heat-resistant adhesive is a necessity if the original is to come in contact with hot projection or thermal copy equipment. If letters are heat resistant, it will be indicated on the dry-transfer sheet.

Required

Sheet of dry-transfer letters
Paper, cardboard, clear acetate, or other art surface
Blunt pencil, ballpoint pen, or other burnishing tool

Procedure

1. Remove or fold back protective backing sheet.
2. Draw guidelines on the art surface, position determined by guidelines provided on transfer sheet. Some brands provide alignment marks above the letters, some below. Some do not provide marks so the lower edge of the letter is aligned on guideline.
3. Position the letters so that the guidelines provided on the transfer sheet and the material to be lettered are aligned.
4. Transfer desired letter by rubbing over it lightly with a dull pencil or a burnishing tool. Be sure to rub over all edges and thin lines to assure total transfer of each letter.
5. Lift the transfer sheet carefully away from lettering surface. Letter should remain fixed in desired position. In case of error, letters can easily be removed with masking tape or a blade, depending on the art surface.
6. Place backing sheet over letter and reburnish carefully.
7. Carefully erase guidelines.

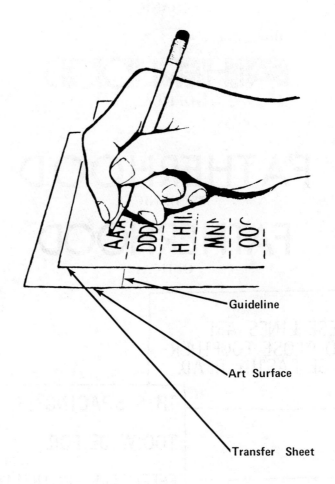

Cardboard Stencils

One way to produce lettering for graphic materials is with cardboard stencils. They provide a fast, easy, economical way to letter and are made of cleanly cut letters on a durable, treated stencil board. A variety of styles and sizes (1/2" to 6") are available. Quality varies however, so you should examine the stencils before purchasing them.

Required

Cardboard stencil
Paper or cardboard
Pencil and ruler

Procedure

1. Draw a light pencil guideline that will be slightly below (or above depending on the brand of stencil) the letters to be traced. Guidelines should be spaced according to the size of the stencil.
2. Before each letter is traced, position guide so that the guideline appears in the center of both guide holes.
3. Trace the outline of the letter with a pencil. The letter can be filled in later with a felt-tip pen or other marking device. If stencil brushes or spray paints are used, it is not necessary to trace the outline of each letter.
4. Before moving the guide, make a spacing reference mark by lightly tracing the guide hole at the right of the letter you have just traced.
5. Position the next letter along the guideline so that the circle you have just made shows through the guide hole at the left of the letter you are going to trace and the guideline is centered in the guide hole at the right. (Adjust letters to the left or right to obtain optical spacing if desired.)
6. Trace the letter and repeat steps 4 and 5 until the word or line has been completed.
7. Erase the guidelines. Finished letters can be filled in with crayons, felt-tip pens, ink, or paints.
8. With a little care you can fill in the gaps left by the stencil. This creates a much more legible and attractive letter.

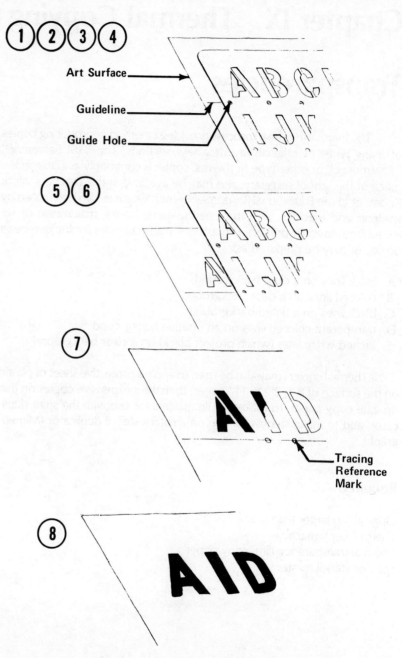

Chapter IX   Thermal Copying

Transparencies

The thermal (dry-heat) process provides a method of producing copies of many types of originals in just a few seconds. The "3M Secretary," "Thermofax," or other type of thermal copier is commonly available in the office of the school secretary, and may be available in the Teachers Work Room, or in the Educational Resources Center. Originals can be created by students and teachers; torn from newspapers, books, magazines; or selected from specially prepared "masters." Transparencies for the overhead projector may be prepared as:

A. black lines on a clear background
B. colored lines on a clear background
C. black lines on a tinted background
D. transparent colored lines on an opaque background
E. etched white lines (which project black) on a clear background

Thermal copiers may also be used to apply a protective sheet of plastic on the surface of 8 1/2" × 11" pages, to make inexpensive copies on thin opaque copy paper, to make purple masters for use with the spirit duplicator, and to make stencils for use on a rotary stencil duplicator (Mimeograph).

Required

Original (no larger than 8 1/2" × 11")
Thermal copy machine
Thermal transparency film, copy paper
Spirit or stencil master

Procedure

1. Select or create an original. The thermal process will reproduce only those configurations that will absorb heat. Pictures with dark *lines* (containing adequate carbon content) copy well; light lines and areas and dark areas do not.

   NOTE: Some copy that is not reproducible by the thermal process (color, faded originals, felt marker, etc.) may be copied on the Xerox first. Then it can be copied by the thermal process, as Xerox lines have adequate carbon content to absorb sufficient infrared heat. Pencil and India ink are also reproducible by the thermal process.

TEARSHEET

2. Edit if desirable. (You may wish to eliminate unwanted material by cutting it away. Additional information may be added by splicing it in, or by writing or typing on the original. Fasten edited copy on plain white paper with transparent plastic or cellophane tape.)
3. Select desired thermal material and place it on top of the original. Typically, the notched or cut corner should be in the upper right-hand corner position as you insert the material into the copy machine. This ensures direct contact of the treated side of thermal material and original. (Specific instructions for each type of thermal material will follow.) It is generally helpful to use a plain piece of paper as a backing sheet for the original. If the original is thin and has dark printing on the back, using a dark-colored backing sheet will reduce "print-through."
4. Plug in the power cord. Set the exposure dial at the position recommended for the type of thermal material you are using. (This will be posted on the machine, as it varies from machine to machine.) See "How to Save Some Money by Using Test Strips" on next page.
5. Insert the material face-up in the machine. (If it does not feed in automatically, check to see if there is an on/off switch on the machine.) CAUTION: NEVER INSERT RIGID MATERIAL (such as cardboard) or STAPLES or PAPER CLIPS in the machine).
6. Separate the copy from the original.

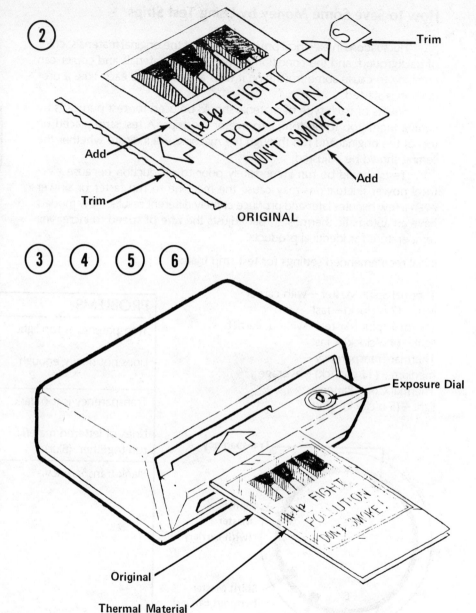

How to Save Some Money by Using Test Strips

Fluctuations in electrical power, density of the original materials, color of background, and the condition of the thermal materials and copier can combine to cause some frustration if tests are not made each time a production is started.

Sheets of each type of material should be used for test purposes by cutting them (top to bottom) into five or six strips. A test strip placed on top of the original and run through the machine will indicate whether the setting should be adjusted.

Tests should be run immediately prior to production because electrical power fluctuations may cause the machine to run faster or slower even a few minutes later and produce entirely different results. New models have an automatic thermostat that adjusts the rate of speed to increasing temperature for identical products.

Initial recommended settings for test strip trials:

Thermal Spirit Master — with carrier
light — (2 o'clock) — fast
Thermal Spirit Master — without carrier
light — (4 o'clock) — fast
Thermal Transparency —
medium — (11 o'clock) — average
Thermal Copy Paper —
light — (5 o'clock) — fast

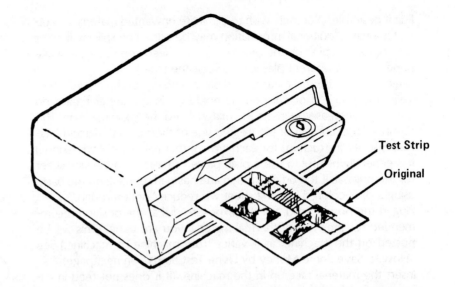

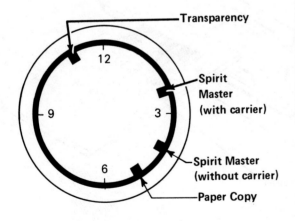

PROBLEMS	POSSIBLE CAUSES	REMEDIES
Transparency is too light. Lines not heavy enough.	Materials moved through machine too rapidly. (Under Exposed)	Set control dial to "darker" setting, which moves materials at a slower speed allowing more time for heat absorption.
Transparency is too dark. Lines of lettered material run together. (Blur)	Materials moved through machine too slowly. (Over Exposed)	Set control dial to "lighter" setting, which moves materials at a faster speed.
Blank transparency.	Lines on the original were not reproducible.	Use carbon-based lead or ink, or make a Xerox copy of the original and use it as an original.
	Inadequate exposure.	Set control to a "darker" setting and rerun the same film.
	Incorrect orientation of materials.	Check instructions to be sure original and thermal material are assembled and inserted correctly.

Thermal Spirit Master Preparation

Most schools have a thermal copier and spirit duplicator. This combination of duplicating machines is advantageous because they produce relatively inexpensive copies that might not otherwise be available. Elaborate, detailed, and professional-looking copies may be reproduced by the spirit duplicator, in combination with the thermal copier.

Required

Thermal copier
Plastic carrier
Selected original
Thermal spirit master

Procedure

1. Select an original that you wish to duplicate or distribute to others. (Because thermal spirit masters will not produce as many copies as direct spirit masters, their use is recommended only when it would save considerable preparation time or provide quality that could not otherwise be attained.)
2. Remove and discard lightweight separation tissue from thermal spirit master. Insert the original between heavy white or yellow backing sheet and glossy side of dye sheet. (The face of original should be in contact with glossy side of dye sheet.) If materials you purchased do not have backing sheet, use a piece of plain white paper in its place.

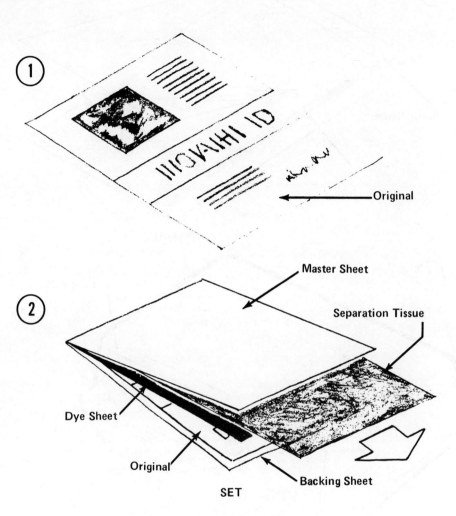

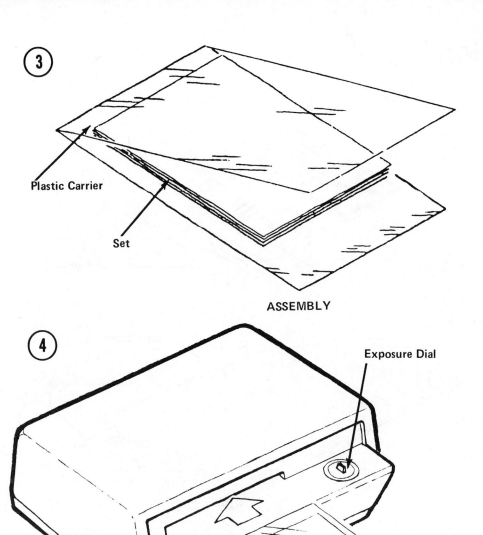

3. Insert the set, comprised of the master sheet, the dye sheet, the original, and the backing sheet, in a plastic carrier. If plastic carrier is unavailable, adjust dial to faster speed. Original should always be inserted into the machine in the faceup position. Use of test strips is suggested but extreme care must be exercised to ensure that dye sheet does not extend beyond master sheet because dye will melt onto plastic carrier or carrier belt, thereby affecting subsequent projects.
4. Adjust the thermal copier to the appropriate temperature. Insert the assembly.
5. The top page of the set will be used as the master to reproduce copies on the spirit duplicator. Extraneous dye deposits along the edges of the master sheet may be trimmed away before duplicating. Unwanted dye within the text may be eliminated by covering with cellophane tape.

Chapter X Spirit Duplication

Quantity Duplication

The spirit duplication process is one of the least expensive means of creating multiple copies of instructional materials. The process involves the transfer of an aniline dye from the back side of a master sheet. Paper that is dampened with duplicator fluid picks up some of the dye as it passes through the machine. A master will produce up to several hundred copies, depending on the skill of the operator and the color and brand of the dye sheet. "Ditto" is a specific brand of spirit duplicating materials and equipment, but the term is often incorrectly used to refer to the process. Spirit duplicators are a common fixture in teacher work rooms and in school offices.

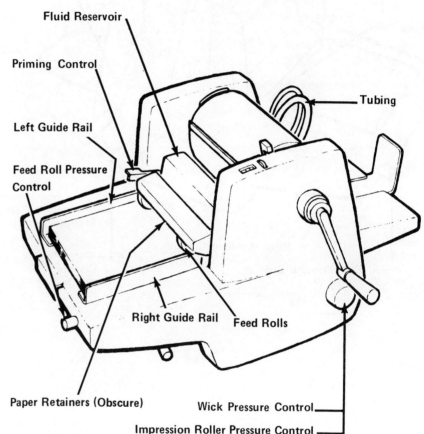

Required

Spirit master to be reproduced
Spirit duplicator
Clean duplicator paper
Spirit duplicating fluid

Procedure

1. Place fluid tank in bracket with spout down into fluid reservoir or pump priming control if reservoir is filled by tubing directly from supply can.
2. Set impression roller pressure control and wick pressure control to operating position (first or second notch).
3. Depress paper loading control.
4. Move left guide rail to 8 1/2.
5. Place stack of paper on feed table against left guide rail.
6. Move right guide rail up to paper stack.
7. Slide paper into machine until stopped by paper retainers.
8. Move feed rolls approximately 1/2" inside guide rails.

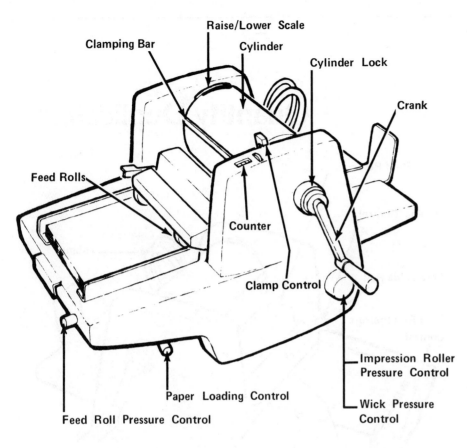

9. Drop feed rolls using paper loading control or press feed roll release control if present.
10. Set feed roll pressure control according to weight of paper. Lightweight paper requires low pressure — heavy paper and card stock require high pressure.
11. Move clamp control to open, turn crank to 6 o'clock position.
12. Place upper edge of master under clamping bar so that dye side of master is to your left (toward feed table). Generally use 8 1/2" markings.
13. Move clamp control to closed.
14. Set counter to 0.
15. Turn crank counterclockwise to print a few copies.
16. If copy is too high or low on the paper: Move impression roller pressure control to off. Loosen cylinder lock. When scribed mark on cylinder is at center line on raise/lower scale, control is in center position. Manually rotate cylinder for appropriate adjustment. Tighten cylinder lock and return impression roller pressure control to correct pressure setting.
17. Turn crank counterclockwise to produce desired number of copies.

Preparation for Idle Periods or Overnight Storage

18. Lift fluid tank and replace in bracket with the spout up. Tube fed systems require no action.
19. Release the wick pressure control.
20. Move impression roller pressure control to off.
21. Depress paper loading control.
22. Move feed roll pressure control to 0.

NOTE: FAILURE TO DO ITEMS 18–22 WILL CAUSE PERMANENT DAMAGE TO THE MACHINE.

Direct Spirit Master Preparation

The spirit duplication process requires that a spirit master be prepared. This process sheet will provide basic information on how to prepare that master.

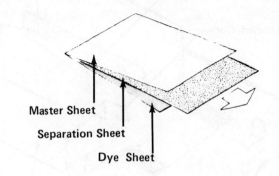

Required

Spirit master
Typewriter
Sharp blade
Ball point pen

Procedure

Preparing the Master

1. Remove separation sheet from between white master sheet and dye sheet.
2. Write, draw, or type on white master sheet so that dye will be deposited on back of master sheet. Even, firm pressure, such as ball point pen or medium hard pencil is desirable for line drawings. Typewriter characters should be cleaned before a master is typed.
3. When finished, separate master sheet from dye sheet. Discard dye sheet to avoid smearing dye on other sheets. If the master is not to be used immediately, restore the original separation sheet next to dye sheet.

Corrections

1. Errors may be corrected while master and dye sheets are in typewriter. They are easier to correct if master and dye sheets are separated at top edge after being rolled into typewriter.
   A. Roll master sheet out far enough so that you can easily reach the spot on the back of master where error occurred. Scrape deposited dye from master sheet at point of error. A sharp knife or razor blade works effectively.

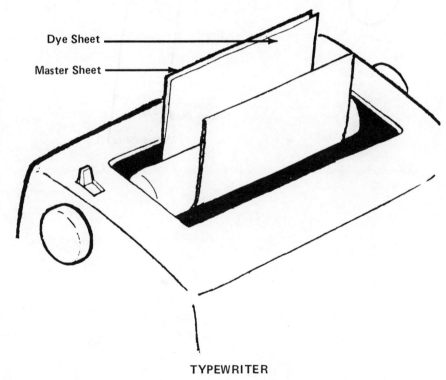

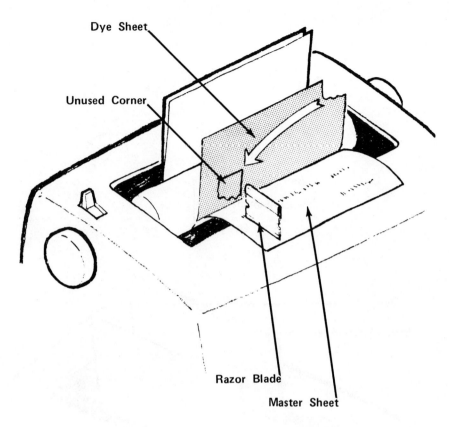

B. Cut or tear a small piece of dye sheet from an unused corner that will not be needed. Place it behind master sheet so it will cover area to be corrected.
C. Retype correction on face of master sheet.
D. Remove the extra piece of dye sheet.
2. Whole sentences may be removed by cutting them out and splicing the master together with transparent tape. Inserts may be made in the same way. Do not put tape on dye side of the master sheet unless you wish to obliterate several lines by covering them with tape.

Multi-Color

Multiple colors may be added on master to emphasize or differentiate selected areas on duplicated copies. Replace original dye sheet with a dye sheet of color you wish to add. Type or draw areas selected for additional color on face of original master sheet.

Chapter XI Materials Preservation

Dry Mounting

Dry mounting is a process by which display materials (pictures, tearsheets, photographs, maps, posters, etc.) may be adhered to backing materials (cardboard, masonite, etc.). Dry mounting is fast, clean, and permanent. It is not necessary to wait for materials to dry or to become involved with messy glue or cement. Dry-mount tissue is a thin sheet of paper coated on both sides with a heat-sensitive adhesive. In use, the tissue is placed between the visual and the backing material. Upon application of heat and pressure, the adhesive softens and adheres the visual to the backing material. Dry mounting may be done easily and quickly with a dry-mount press or with a household iron.

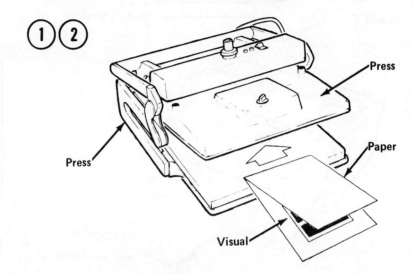

Required

Visual to be mounted
Backing material
Dry-mount tissue
Paper cutter or sharp blade
Dry-mount press and a tacking iron or household iron
Large sheet of clean paper

Procedure

1. Plug in the dry-mount press and the tacking iron. Set the thermostatic control on the press at the temperature appropriate for the mounting material being used. (See chart A.) Set tacking iron at Hi.
2. Place the visual and backing material between clean sheets of paper for protection. This paper will prevent adhesive from contacting the heating element or cushion pad and protect the material from adhesive and dirt that might be on the platen. Insert the visual and backing material in the press for about thirty seconds to remove moisture. Close, but do not lock, the press. The materials must be flat and completely dry.

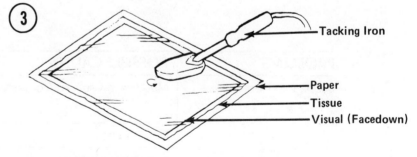

3. Place visual facedown on a piece of clean paper. Tack a sheet of dry-mounting tissue, slightly larger than the visual to the center of the back of the visual by a quick circular motion with the tip of the tacking iron on the surface of the tissue.
4. Turn assembly faceup and trim visual and the tissue with a sharp blade or paper cutter.

183

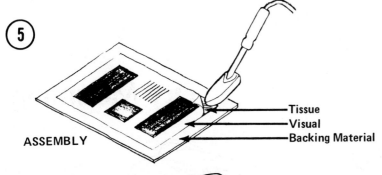

ASSEMBLY — Tissue, Visual, Backing Material

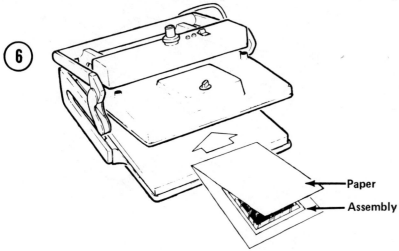

Paper, Assembly

5. Position visual and tissue on the backing material. Lift corners of visual and tack two opposite corners of the tissue to the backing material.
6. Envelop assembly in a piece of clean paper and place in the press for ten or fifteen seconds. (Additional time is sometimes required.) Open press to release moisture, then reheat.
7. Remove from press, place on clean surface to cool under weight to minimize warping. Metal weights speed the cooling process.
8. After cooling, mounted visuals can be checked for adherence by bending slightly toward face. If bubbles or creases can be seen, return project to press until adherence is satisfactory.

MOUNTING MATERIAL	TEMPERATURE
COLOR SEAL	200 degrees
FOTOFLAT	200 degrees
CHARTEX	200 degrees
MT5	225 degrees
KODAK DRY-MOUNT TISSUE	225 degrees

Chart A

PROBLEMS	POSSIBLE CAUSES	REMEDIES
Bubbles and wrinkles.	Moisture in visual and/or backing material.	Preheat unmounted materials.
		Replace mounted materials in press for several minutes.
	Press too hot.	Cool press to a lower temperature.
	Improper tacking.	Replace in the press. (May diminish but not disappear.)
Nonadhesion.	Press too cool.	Increase press temperature, reheat materials.

Laminating

The lamination process may be used to protect maps, pictures, and tearsheets from dirt and moisture. The process is not complicated, messy, or difficult. Laminating film is tough, mylar material that has been coated on one side with a heat-sensitive adhesive. Upon application of high temperatures and pressure, the heat-sensitive adhesive softens and adheres to most clean surfaces. Some items, such as color photographs, can be damaged by high temperature required.

Required

Visual to be laminated (mounted or unmounted)
Laminating film
Paper cutter or sharp blade
Dry-mount press and a tacking iron or household iron
Large sheet of clean paper

Procedure

1. Plug in dry-mount press and tacking iron. Flip switch on. Set thermostatic control on press at 270. Set tacking iron at Hi.
2. Preheat materials to be laminated to remove moisture. Insert in press for about 30 seconds with press closed but not locked. Unmounted visuals or visuals that have been dry-mounted to cardboard can be laminated.
3. Remove materials from press and cover with slightly larger piece of laminating film. Dull side has heat-sensitive adhesive and should be in contact with surface to be laminated. If project is likely to be subjected to spills, both surfaces should be laminated; for most applications, lamination of just face is sufficient. Static electricity created by smoothing film with your hands will generally hold film in position. Sometimes you may wish to fold film loosely around edges and tack on the back with a tacking iron.

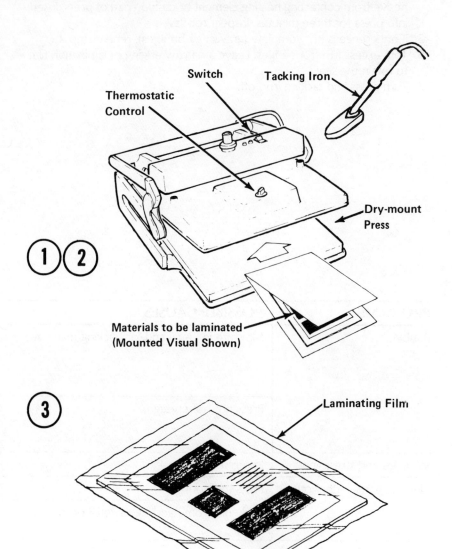

Assembly
(Mounted Visual)

4. Place assembly in protective envelope of clean paper to prevent adhesive from contacting heating element or cushion pad of press. Insert into press for three minutes, inspect for flaws.
5. If adherence is not complete (grayish air bubbles), repeat step 4.
6. Trim excess film from edges. Leave a narrow edging of lamination film to retain the seal.
7. Turn press and tacking iron **off.**

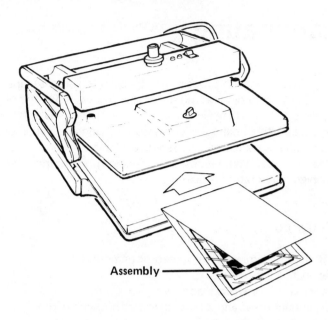

Assembly

PROBLEMS	POSSIBLE CAUSES	REMEDIES
Bubbles	Moisture in visual and/or backing material.	Preheat unmounted materials. Bubbles may be pierced to remove moisture. Replace mounted materials in press for several minutes to one hour.
	Press not hot enough.	Allow press to gain correct temperature. Thermostatic indicator will go out when selected temperature is achieved.
Wrinkles and stress lines	Improper tacking.	Laminating film should not be under stress when tacked.
Non- or partial adhesion	Press too cool.	Increase temperature, reheat materials.
	Body oils from handling material or moisture.	Increase temperature and pressure, reheat materials. Increase time in press. Keep work clean in the future.

Chapter XII   The Copyright Law and Its Implications

Limitations of This Chapter

This chapter is limited to a short overview of copyright law at a particular moment in time. It expresses the view of an educator who teaches copyright basics to students of instructional design and technology and to students of library and information science. There is no intent to provide legal advice to any reader. Individual teachers, instructional designers, and librarians should seek legal counsel for specific copyright questions.

Definition

Copyright is a legal protection that grants an author or creator the exclusive right to control the reproduction, distribution, performance, and public display of a copyrighted work, as well as the right to the control of derivative works.

What Can Be Copyrighted

Literary works, musical works, dramatic works, pantomimes, and choreographic works; pictorial, graphic, and sculptural works; motion pictures, including video; sound recordings and other audiovisual works. Computer programs qualify as literary works and are protected even if recorded on a silicon chip.

International Copyright Protection

There is no universal international copyright agreement. The extent of protection and the requirements for securing copyright vary from country to country. Many countries are party to one or more international copyright treaties or conventions.

U.S. Copyright Law

A new United States copyright law (U.S. Public Law 94-553) took effect on January 1, 1978. The law brought the duration of copyright in line with international practice and provided for media formats not envisioned at the time the 1909 law was passed.

In the United States, copyright is administered through the Copyright Office of the Library of Congress (address at the end of the chapter).

Copyright Notice

The notice of copyright includes three components:

1. the copyright symbol "©," the word "copyright," or the abbreviation "copr.," or "P" in the case of phonograph records
2. the year of first publication
3. the name of the copyright owner

Works published prior to 1978 forfeited protection if the copyright notice was omitted. Unpublished works were not protected under the 1909 copyright law although some protections existed under common law.

As of January 1, 1978, copyright attaches to a work from the time it is fixed in tangible form; i.e., written down or otherwise recorded. An unpublished work created on or after that date receives protection even if it does not carry a copyright notice. Published material must contain a proper copyright notice or under certain circumstances protection may be forfeited.

Copyright Registration

The copyright owner deposits one copy of an unpublished work or two copies of a published work at the Copyright Office. A completed registration form and a nominal fee accompany deposit copies. Forms for registration of a copyrighted work may be obtained from the Copyright Office.

Although the 1978 law does not *require* registration of a claim, copyright/registration is necessary before any infringement suits can be filed in court. A work may be registered at any time within the life of the copyright.

Duration of Copyright

Under the current law, copyrighted works receive protection during the author's lifetime plus 50 years. There is no renewal of copyright for works created after January 1, 1978.

Under the 1909 law, published works were protected upon registration for a term of 28 years with the option of a further 28 years if renewed. However, the majority of copyrights were not renewed.

Public Domain

When a copyright term runs out, the material is considered to be in the public domain and may be freely copied and/or performed.

For works first published before 1978, absence of a copyright notice may indicate that the work is in the public domain.

Most U.S. government documents are not eligible for copyright protection and are in the public domain upon creation if they have been created by employees of the U.S. government in the course of their employment. Works done under contract to the government may or may not be copyrighted according to the terms of the contract. State and local governments are not restricted from claiming copyrights for their documents.

Determining if a Work Is in the Public Domain

Works Published before January 1, 1978

Works first published over 75 years ago may be assumed to be in the public domain. Works published before 1978 that lack a copyright notice are also in the public domain.

Works First Published less than 75 Years Ago and Prior to January 1, 1978

To be certain of copyright status, first check for the date of the copyright notice on the work. Then pay to have a search done at the Copyright Office. There is a nominal fee, payable in advance. Write to the Copyright Office for a circular explaining current practice.

Published and Unpublished Works Created after January 1, 1978

Works created after January 1, 1978 (whether published or unpublished) are protected for the life of the creator plus 50 years. Protection is not forfeited if the copyright notice is omitted.

Unpublished Works Created before January 1, 1978

Works that were not published or copyrighted before January 1, 1978 are protected for the life of the author plus 50 years, or at least until December 31, 2002. If such works are published before 2002, then they will be protected until December 31, 2027.

Fair Use

Although copyright grants certain exclusive rights to the owner of the copyright, the new copyright law also makes provision for "fair use" of copyrighted material by others. Fair use permits limited copying of material without permission from, or payment to, the copyright owner.

Section 107 of P.L. 94–553 sets four tests of fair use:

1. the purpose and character of the use (only nonprofit educational uses are considered to be "fair")
2. the nature of the copyrighted work

3. the amount of the work used in relation to the whole
4. the effect of the use on the potential market of the work

To be considered "fair use" a particular instance generally must pass all four tests.

Guidelines Defining Fair Use

Several sets of guidelines for defining fair use have been negotiated by groups of educators, publishers, and authors: "Agreement on Guidelines for Classroom Copying in Not-for-Profit Educational Institutions"; "Guidelines for Educational Uses of Music"; "Guidelines for Off-Air Recording of Broadcast Programming for Educational Purposes." These guidelines do not have the force of law, but they may be used by a court to interpret the law in a specific instance. The various guidelines have been widely reprinted and appear in their entirety in several of the documents cited in the references at the end of the chapter.

The guidelines should be considered to set minimum standards rather than indicating the maximum uses permissible. In other words, you might legally be able to do more, but you will not know how much more is safe because the "more" has not been defined yet.

The following sections provide an overview of fair use for specific media formats but do not cover every situation included in the guideline documents or each court decision that has interpreted fair use. Uses considered fair for one media format may be prohibited for a different media format.

Fair Use of Printed Material

Individual Use

An individual may make one copy of the following for his or her own research or teaching; an article, a book chapter, a short story, a short essay, a poem, or an illustration from a book, periodical, or newspaper.

Classroom Use

Multiple copies must be made by or at the specific request of the teacher of a course. No more than one copy per pupil may be made. Each copy must include a copyright notice. Copying must not substitute for purchase of materials. Consumable materials such as workbooks, tests, and answer sheets may not be copied. Tests of brevity, spontaneity and cumulative effects must be met.

Brevity

Poetry: No more than 250 words or 2 pages, whether an entire poem or an excerpt from a longer poem.

Prose: A complete story, article, or essay of less than 2,500 words, or an excerpt not to exceed 1,000 words or 10% of a work, whichever is less. Picture books and similar shorter works combining short text and illustrations may not be reproduced in their entirety (whether through photocopy or other photographic means such as slides or video). An excerpt may be reproduced if it comprises no more than two pages and contains no more than 10% of the words in the text.

Illustrations: A single chart, graph, diagram, drawing, cartoon, or picture from a particular book or periodical issue may be reproduced. Multiple copies of a particular cartoon may not be made.

Spontaneity

The decision to use the material and the date of use must be so close in time that permission of the copyright holder could not be sought and received prior to the use of the material in instruction.

Cumulative Effect

The copying may be done for only one course in the school. No more than nine instances of multiple copying may take place for one course during a class term. No more than one story, article, essay, or poem or two excerpts by the same author may be copied during a single term. No more than three excerpts may be copied from a collective work or one volume of a periodical. An exception to these limits is made for current news publications.

Fair Use of Musical Scores and Arrangements

Performance

Emergency copying may be done to replace lost or damaged purchased copies when needed for an imminent performance provided that new purchased copies are subsequently acquired and the temporary copies are destroyed.

Academic Purposes other than Performance

Multiple copies of excerpts of musical works may be made provided they do not exceed 10% of the whole work and provided that they do not consist of an entire performable unit such as an aria or movement. No more than one copy per pupil may be made.

A single copy of an entire performable unit such as an aria or movement may be made by or for a teacher for his or her use in scholarly research or in preparation for teaching, provided that the copyright holder confirms that the work is either out-of-print or unavailable except as part of a larger work.

Purchased printed copies may be simplified or edited, provided that the fundamental character of the work is not distorted. Lyrics may neither be added nor altered.

Fair Use of Audio Materials (On Record, Disc, Tape, or Radio)

One copy of a portion of a copyrighted sound recording or radio program may be made by a student or by others for student use. The copy cannot be sold or used outside of the educational context in which it was made.

A single copy of a portion or an entire musical sound recording may be made from a recording owned by a teacher or an educational institution if used for instruction or for examination. The teacher or institution may retain such copies.

A single copy of student performances of a copyrighted work may be made for study or for evaluation. The copy may be retained by a teacher or an institution.

Copies of National Public Radio broadcasts may be made and retained by schools. Copying of commercial radio broadcasts follows the same rules that apply to off-air taping of commercial television programs (see below) except for copyrighted music.

Request permission of copyright owners if multiple copying or creation of anthologies or compilations of copyrighted audio material are contemplated.

Fair Use of Slides, Filmstrips, Films, and Prerecorded Video

One copy of a *portion* of a copyrighted slide series, filmstrip, prerecorded video, or film may be made by a student or by others for a student provided the material is owned by the institution.

A single copy of a portion (but not the whole) of a slide series, filmstrip, prerecorded video, or film may be made by a teacher for scholarly or teaching purposes. Care must be taken to avoid copying the "creative essence" of a work, even if what is copied is only a very small portion of the whole.

Materials sent for preview may not be copied under any circumstances. A film sent for preview may be shown via closed-circuit television, but only within the immediate building.

Fair Use of Video Material

Commercially Distributed Videotapes

When video material is acquired (through sale or lease), the purchase does not include permission to make additional copies, e.g., archival copies. Users may not transfer the material from one format to another, e.g., make a 1/2" VHS videocassette copy of a 3/4" video recording.

Commercial film and video distributors provide their materials either for "Home Use Only" or cleared (licensed) for public performance. The type of use permitted to a purchaser will vary depending on the rights purchased with the material when it is sold or leased.

Materials licensed or sold for "Home Use Only" are to be used by individuals in their homes with their families or social acquaintances. Such materials may not be presented in any type of public performance regardless of whether the performance is for the benefit of a for-profit or a not-for-profit organization or group.

Materials licensed or sold for "Home Use Only" may be used by nonprofit educational institutions so long as the use is in face-to-face teaching in a classroom or similar place devoted to instruction; recreational or entertainment uses are not permitted. Use in a library-based educational program probably would not be infringing if the face-to-face instructional use requirements are met.

Libraries may circulate copies of prerecorded video materials to patrons for use in their homes. However, the libraries may not show the same video materials in a public performance unless they have negotiated the right to do so as part of a licensing agreement or unless they have sought and received permission to do so from the copyright owner. Private viewing in a library probably qualifies as home use.

Off-air Recording of Broadcast Programming for Educational Purposes

Nonprofit educational institutions may record off-air television broadcasts (including cable television retransmission) and retain a copy for a period of up to 45 consecutive calendar days after the recording date. After this time, the recording must be either erased or destroyed.

Such a recording may be used once by an individual teacher in the course of relevant teaching activities. It may be shown one additional time if necessary for reinforcement of student learning. Such use must take place during the first 10 consecutive school days within the 45 calendar day retention period. After the 10 consecutive school day period, the recording may only be used by the teacher for evaluation purposes.

Recordings cannot be made in anticipation of teacher requests. They must be made at the specific request of an individual teacher for the use of the requesting teacher. A program cannot be off-air recorded for a teacher more than once even if it is broadcast at additional times.

The recorded programs may not be altered from their original content; however, they need not be used in their entirety. The copyright notice must be included on each copy of an off-air recording as given on the broadcast program.

Each educational institution is expected to establish and carry out appropriate reporting and control measures to insure compliance with the guidelines for off-air videotaping.

Although a 1984 Supreme Court ruling established that individuals may record any television program off-air at home for later viewing, the ruling applied only to personal home use. Such copies cannot be made by an instructor for later use in an instructional setting.

Fair Use of Computer Software

In 1980 Congress amended the Copyright Act to include a specific section on computer software. At this time there are no negotiated guideline agreements further defining fair use. To determine fair use of computer software, you must know whether the software has been sold outright or purchased under a license or lease agreement.

Licensing

Most computer software is not sold outright. The package may contain a printed agreement or contract indicating the terms of sale. Frequently this agreement is placed under a clear plastic shrink-wrap and contains a clause stating that removal of the shrink-wrap by the purchaser constitutes agreement to the terms of the contract.

The terms of license agreements may vary widely even within a single company's product line. Use on more than one computer during the same time period or loan of the material by libraries may be prohibited. Libraries and schools may wish to modify license agreements. A phone call to the publisher can determine if alternate licensing agreements are available. Some institutions include a statement on their purchase orders indicating that the order should not be filled if the material cannot be loaned for use at home or in the library.

Archival Copying

The owner of a computer program may make a single copy for archival purposes. The archival copy must contain the copyright notice. Either the archival copy or the original copy may be used; however, both copies may not be used at the same time. Another archival copy may be made if the circulating copy is damaged or destroyed.

If a computer program is resold, the archival copy must be transferred with the original program or destroyed.

Instructional and Library Use

Observe any licensing restrictions.

If a single copy of licensed software is owned, use it on one machine at a time. Loading one purchased copy into several machines for use at the same time or using it in a network of computers would probably be viewed as an attempt to avoid purchase of multiple copies.

Post warnings on machines used by students and patrons that indicate that unauthorized copying of copyrighted programs is prohibited.

Penalties for Copyright Infringement

The law provides for infringement penalties ranging from $250 to $50,000 and/or up to one year's imprisonment for *each* violation. Even if a defendant is judged not guilty of infringement, court fees and attorney's costs must be paid.

Requesting Permission to Copy

If a work is not in the public domain, and you have reason to believe the use contemplated may not be permitted under the fair use provisions of the law, you may still legally copy or perform the work if you seek and receive permission to do so from the copyright owner. Write to the author or publisher specifying exactly what you wish to do and why. A sample letter form is provided at the end of this chapter. To locate current addresses for publishers, consult *Literary Marketplace,* a publication of the R. R. Bowker Company, which is available in most libraries.

The permission must be given explicitly. You cannot assume permission was given if your request does not result in a reply. An organization called the Copyright Clearance Center (CCC) handles permissions and collects fees for a group of publishers. Write for a list of the publications covered by their service. Corporations and libraries may arrange to pay an annual fee based on the amount of copying done. The Television Licensing Center provides information and services for those seeking to copy video materials. Addresses for both organizations are given at the end of this chapter.

Institutional Policies

Many institutions have developed copyright policy documents that inform their employees of the fair use guidelines. Through the policy document, individuals may also be informed that an employee who violates the law will not receive institutional support in the event of a court case.

Changes to the Law

The Copyright Office is required to review portions of the law at five-year intervals and to submit a report of the review to Congress. The review must address the problems encountered in balancing the rights of creators and the rights of users. The Copyright Office is further required to propose legislative changes or other recommendations if warranted.

If you would like to see changes in the law, write to the Copyright Office and/or your congressional representatives. Identify problems, propose changes, and provide the rationale for your recommendations.

Keeping Current

Copyright is a fast-moving area. Technological advances raise new questions. Creator and user groups lobby for changes to the Copyright Act. Amendments to the Copyright Act are proposed and may become law. Court decisions provide new interpretations of the law and its amendments. Some of what you read in this chapter may be out-of-date by the time of publication. Anyone likely to be affected by changes in the law needs to develop a habit of scanning newspapers and professional journals to keep up with current events in this area.

Responsibilities of Media Professionals Regarding Copyright

To uphold the law and to follow the guidelines.

To inform others of the provisions of the law and the guidelines.

To follow court decisions and/or proposed changes to the law and guidelines.

To lobby for changes and improvements to the law and guidelines.

Addresses

Copyright Clearance Center
310 Madison Avenue
New York, NY 10017

Copyright Office
Library of Congress
Washington, D.C. 20559

Television Licensing Center
5547 N. Ravenswood Avenue
Chicago, IL 60640

Selected Copyright Resources

Copyright Office Information Kit. Copyright Office, Washington, D.C., 20559.

"Current Problems in Copyright," *Library Trends* 32:2 (Fall, 1983) Theme issue. Walter C. Allen and Jerome K. Miller, issue editors.

Johnson, Donald F. *Copyright Handbook.* 2nd ed. New York: R. R. Bowker, 1978.

Miller, Jerome K. *Using Copyrighted Videocassettes in Classrooms and Libraries.* Champaign, IL: Copyright Information Services, 1984.

Nonprint Media and the Copyright Law: An Educator's Responsibilities and Rights. [pamphlet] Compiled by A. Brian Helm, Director of Library Media Services, Anne Arundel County Public Schools. Reprinted with Permission by the Iowa Dept. of Public Instruction, 1985.

Reed, Mary Hutchings. *The Copyright Primer for Librarians and Educators.* Chicago: American Library Association; Washington, D.C.: National Education Association, 1987.

Stanek, Debra. "Videotapes, Computer Programs and the Library." *Information Technology and Libraries.* 5:1 (March, 1986), pp. 42-54.

Troost, F. William. "Students — The Forgotten People in Copyright Considerations." *EITV* 15:6 (June, 1983), pp. 70, 72-74.

Example of letter to request permission to duplicate copyrighted material.

(Your letterhead at top of form)

Dear Friends:

We request permission to duplicate the following material for use in our course, _____ .   *Indicate if you are requesting one-time or continuous use.*

Title:
Author:
Publisher:
Copyright date:

Material to be duplicated: *Give page numbers and/or chapter numbers.*
If continuous, number per semester.

Number of copies needed:

Distributed through:

Type of reprint: *E.g., photocopy.*

Type of use: *E.g., required reading; supplementary reading.*

The price charged to students will include only the cost of duplication.

We would appreciate your reply to our request by _____ , so that we can complete arrangements for classroom instruction.   *Allow two weeks. After second week, send follow-up request.*

A self-addressed stamped envelope and a copyright permission form are enclosed for your convenience. Please advise us if there will be a fee for your permission to duplicate. Funds available for copyright fees are limited.   *Permission form repeats bibliographic information; contains a signature line and a date line.*

We appreciate your consideration in providing materials to assist student learning.

Sincerely,

(title of requestor)

APPENDIX

Study Questions

Projection (General)

1. Films, filmstrips, and slides are inserted in projectors in head-down positions.
   a. True
   b. False

2. Keystone effect is the term applied to a distortion of picture shape caused by the angle of projection.
   a. True
   b. False

3. At a given distance a three-inch lens will project a larger picture than a five-inch lens.
   a. True
   b. False

4. An overhead projector is a good example of a reflected light system.
   a. True
   b. False

5. An opaque projector is a good example of a direct transmitted light system.
   a. True
   b. False

6. Because film moves through a motion picture projector top to bottom, dirt and emulsion will collect on the upper edges of the pressure plate and aperture plate.
   a. True
   b. False

7. A keystone effect is caused by projected light rays falling on one part of the screen having to travel a greater distance than those falling on another part of the screen.
   a. True
   b. False

8. The distance from screen to the last row of students is 24 feet. What should be the minimum width of a wall-mounted screen if the last row of students is to see properly?
   a. 4'
   b. 6'
   c. 8'
   d. 9'

9. Vertical keystoning of an image is reduced by:
   a. tilting the top of the screen forward or the bottom backward
   b. elevating the projected image
   c. moving the projector closer to the screen
   d. a and b

10. Select the most accurate statement.
    a. After prefocusing a projector, further focusing is not necessary.
    b. Prefocusing is important because it enables proper projector placement with relation to the screen.

11. Which of the following factors have a major influence on the brightness of a projected image?
    1. distance from projector to screen
    2. ambient light that falls on the screen
    3. size of the room
    4. type of screen
    a. 1, 2, and 3
    b. 2, 3, and 4
    c. 1, 2, and 4
    d. 1, 3, and 4

12. For the most flexibility in projection, select a _____ screen and a _____ lens.
    a. beaded – long focal length
    b. matte – short focal length
    c. lenticular – long focal length
    d. lenticular – zoom
    e. matte – zoom

13. When a projector is moved away from a screen, the picture becomes:
    a. larger but dimmer
    b. smaller but brighter
    c. smaller but dimmer
    d. larger but brighter

14. If students in your classroom are seated at wide angles from the line of projection, which type of screen would you use?
    a. beaded
    b. matte

15. A glass-beaded screen is recommended for classrooms with dimensions that are:
    a. long and narrow
    b. square
    c. both a and b
    d. none of the above

16. Placing the screen in the corner of the room on a window wall instead of in the center of the chalkboard wall has which of the following advantage(s)?
    a. less ambient light falls on the screen resulting in brighter image
    b. less space is required
    c. keystoning is reduced
    d. all of the above
    e. none of the above

17. If the projected image on the screen is too large, it can be reduced by:
    a. moving the projector farther away from the screen
    b. replacing the objective lens with one of a shorter focal length
    c. adjusting the film gate
    d. moving the projector closer to the screen

18. The basic components necessary for a projection system are:
    1. source of light
    2. reflector to control the direction of the light
    3. lens system (which may include mirrors)
    4. means for supporting material to be projected
    5. surface for viewing the projected image
    a. 1, 3, and 5
    b. 2, 4, and 5
    c. 1, 2, and 4
    d. 1, 2, and 5
    e. all of the above

19. To eliminate keystoning, the screen should be placed:
    a. perpendicular to the line of projection
    b. horizontal to the line of projection
    c. closer to the projector
    d. at the same level as the projector

20. When a projection lamp is hot, it should be:
    a. cooled with the fan
    b. allowed to cool slowly and naturally without use of the fan
    c. moved very carefully because a hot projection bulb is extremely vulnerable to filament breakage
    d. a and c
    e. b and c

Overhead Projector

1. Translucent materials usually can be projected.
   a. True
   b. False

2. Which of the following items can be projected on the overhead projector?
   1. cardboard silhouettes
   2. transparent plastic tools, models, and templates
   3. drawings on tracing paper
   4. transparencies produced on the thermal copier
   a. 1, 2, and 3
   b. 1, 3, and 4
   c. 2, 3, and 4
   d. 1, 2, and 4

3. The overhead projector can be used in a lighted room because:
   a. it is used at the front, closer to the screen
   b. it has a brighter bulb
   c. large transparencies allow more light to be transmitted
   d. all of the above

4. Unique characteristics of overhead projectors, when compared with other types of projectors, enable users to:
   1. face the audience while looking at the material being projected
   2. be in a lighted room
   3. write on and manipulate the projectual easily
   4. create motion with a simple attachment
   a. 1, 2, and 3
   b. 2, 3, and 4
   c. 1, 3, and 4
   d. 1, 2, 3, and 4

5. Which of the following are reasons to mount a transparency on a mounting frame that covers the entire stage of the overhead projector?
   1. It holds the transparency flat.
   2. It facilitates handling and storage.
   3. It prevents distracting light around the borders.
   4. It provides a place for lecture notes.
   a. 1, 2, and 3
   b. 2, 3, and 4
   c. 1, 3, and 4
   d. 1, 2, 3, and 4

Opaque Projector

1. Materials to be projected are placed in the opaque projector with the upper edge away from the screen.
   a. True
   b. False

2. The opaque projector is most often used for projection of transparent materials.
   a. True
   b. False

3. The high temperature of the bulb in an opaque projector may cause:
   a. loss of color on the image
   b. distortion of the projected image
   c. heat damage to the software
   d. dangerous gases to form
   e. a, b, and c

4. A disadvantage of the opaque projector is that it:
   a. is difficult to operate
   b. will not project colors
   c. has inefficient light output
   d. is dangerous to use

5. Typically opaque projectors can project entire images that do not exceed:
   a. 8½" × 11"
   b. 9" × 12"
   c. 18" × 18"
   d. 15" × 15"
   e. 10" × 10"

6. One problem with the opaque projector is that it:
   a. frequently crumples material in the belts
   b. requires a relatively dark room
   c. has extremely high voltage
   d. is relatively new and few schools have one

7. All of the following are normally used in an opaque projector except:
   a. flat pictures
   b. three-dimensional objects
   c. 35mm slides
   d. printed maps

Filmstrip Projector with Audiocassette Playback

1. If parts of two frames are projected on the screen, you should adjust the:
   a. filmstrip advance control
   b. framer control
   c. elevation control
   d. all of the above

2. Filmstrips should be inserted in the projector with the image:
   a. head end first and right side up
   b. tail end first and right side up
   c. head end first and upside down
   d. tail end first and upside down

3. The filmstrip may be easily removed from the storage container by:
   a. bouncing it on the palm of the hand to cause it to fall out
   b. pulling on a loose end to remove it from the container
   c. pushing against trailer or inside of strip to make it smaller
   d. none of the above

4. To advance the filmstrip until focus frame appears on the screen, you should:
   a. move film speed control to fast
   b. move off/fan/lamp control to lamp
   c. depress ADV control
   d. do all of the operations listed above

5. Once projector is loaded and advanced to focus frame, the film speed control should be in the:
   a. off position
   b. fast position
   c. single position
   d. normal position

6. The volume control must be on:
   a. only when an audiocassette will be used
   b. to advance filmstrip anytime
   c. only when the play key is depressed
   d. to rewind the filmstrip

7. An audiocassette is loaded by first depressing the:
   a. play key
   b. fast forward key
   c. stop/eject key
   d. rew key

8. At conclusion of program, you should first depress:
   a. stop/eject key only far enough to stop tape
   b. fast forward key
   c. rewind key
   d. reverse key

9. If tones to signal change of visual are heard:
   a. the film speed should be fast
   b. the 50 Hz should be used
   c. the manual control must be used
   d. the volume control should be turned down

10. To lower projector for storage, you should:
    a. turn off the off/fan/lamp control
    b. depress elevation control and push down on front panel
    c. release elevation control
    d. perform b and c above

Multiple-load Slide Projector

1. An advantage of the carousel slide projector is that it can be adapted to project filmstrips.
    a. True
    b. False

2. Both 35mm slides and "super slides" can be projected with a 2×2 slide projector.
    a. True
    b. False

3. When inserting a slide into a projector, the thumbspot should be in the _____ corner.
    a. upper right
    b. lower right
    c. lower left
    d. upper left

4. The lock ring on a carousel tray should be locked into position to:
    a. make it possible to position the tray on the machine
    b. prevent the slides from falling out if the tray is inverted
    c. cause the advance mechanism to engage the tray when activated
    d. engage the remote control

5. The select control on a carousel projector is used to:
    a. release the slide tray for manual advance
    b. allow the user to project a nonchronological, predetermined slide sequence automatically
    c. project the slide in the slot after the one indicated by the gate index
    d. none of the above

6. A slide held so that it "reads" correctly to you should have the thumbspot in the _____ corner.
    a. upper left
    b. lower right
    c. lower left
    d. upper right

Audio (General)

1. Room noises must be kept at a minimum when using a patch cord to record directly from another tape recorder.
    a. True
    b. False

2. IPS is the relative measure of sound intensity or volume.
    a. True
    b. False

3. Placing a microphone farther from the talent usually results in picking up more room noise.
    a. True
    b. False

4. In audio recording the term dubbing means:
    a. using nonmagnetic tape
    b. adding music or other sound after the recording has already been made
    c. making a copy of a tape that has already been recorded
    d. taping live with an open microphone

5. Which of the following is a transducer?
    a. speaker
    b. amplifier
    c. turntable

6. Which of the following affect the playing time of a reel of magnetic tape?
    1. recording speed
    2. thickness of the tape
    3. the frequency range
    4. numbers of tracks recorded on the tape
    a. 1, 2, and 3
    b. 2, 3, and 4
    c. 1, 3, and 4
    d. 1, 2, and 4

7. If you wanted the longest recording possible on a 7" reel audiotape, you should record at _____ inches per second on _____ mil tape.
    a. 1   7/8 — .5
    b. 7   1/2 — 1.0
    c. 3   3/4 — 1.5
    d. 7   1/2 — .5
    e. 1   7/8 — 1.5

8. The number of tracks that can be recorded on a blank audiotape is determined by the type of:
    a. tape used
    b. recorder used
    c. both a and b

9. Which type of recorder allows greater ease of editing audio materials?
    a. cassette
    b. open reel
    c. disk
    d. cartridge

10. When splicing a tape, it is best to cut it at a
    a. 45 degree angle
    b. 60 degree angle
    c. 90 degree angle

11. The recommended method of splicing magnetic recording tape is to:
    a. overlap the ends and apply tape
    b. overlap the ends and apply cement
    c. butt the ends and apply tape

12. The best technique for recording a radio program is to:
    a. place a microphone within one foot of the radio speaker
    b. place a microphone on a pillow at least five feet away from the speaker
    c. use two microphones, one within three feet of the speaker and the other at least ten feet away
    d. connect patch cord to radio speaker terminals or output jack

Audiocassette Tape Recorder

1. When making a tape recording, a standard machine will simultaneously erase any previous recording made on that track.
   a. True
   b. False

2. Automatic level control (ALC) refers to a control of recording volume that will automatically adjust to sounds in range of the microphone.
   a. True
   b. False

3. Batteries should be removed when using AC power.
   a. True
   b. False

4. The safeguard interlock on a cassette tape recorder:
   a. prevents accidental rerecording
   b. automatically adjusts the sound level
   c. prevents the cassette from being removed while the recorder is on
   d. a and b

5. Cassette recorders with sync pulse capability will, when recording a new program:
   a. not erase previous sync signals
   b. erase previous sync signals
   c. record sync signals across all audio tracks on the tape
   d. b and c above

6. Cassette recorders with sync pulse capability can:
   a. automatically control volume
   b. be programmed to accept open-reel tape on 5-inch reels
   c. be programmed to trigger a slide projector
   d. do all of the above

7. When storing a cassette recorder, the function control should always be left in the:
   a. fast forward position
   b. rewind position
   c. stop or neutral position
   d. record position

8. A tape position indicator tells you:
   a. how many feet of tape have gone through the machine
   b. how many feet of tape are left on the cassette
   c. the number of revolutions of the take-up reel
   d. the number of revolutions of the feed reel

9. If there is a switch on the microphone, it indicates that:
   a. it is a wireless microphone
   b. there will be a second plug on the cord that should be inserted into the remote jack
   c. you can control the recorder from a remote position subject to the length of the cord
   d. a and b
   e. b and c

10. It is possible to edit by connecting two recorders with a _____ to transfer selected signals from one tape to a new one.
    a. transport system
    b. speaker
    c. patch cord
    d. transducer

11. When using audio equipment, connect inputs to:
    a. inputs
    b. outputs
    c. speakers
    d. headphones

12. External microphone can be plugged into the Sharp audio recorder through the:
    a. aux input jack
    b. mic output jack
    c. aux output jack
    d. external mic jack

13. To activate the safeguard interlock, you should:
    a. break off tabs on the back of the cassette
    b. put tape over the tabs
    c. leave tabs in place on the cassette
    d. none of the above

14. An omnidirectional microphone is used to:
    a. record group interaction
    b. eliminate background noises in recording
    c. record face-to-face interviews
    d. reduce interference and static

15. Four of the most basic components of sound reproduction equipment are:
    a. sound source, microphone, recording heads, playback
    b. microphone, speaker jack, magnetic tape, erase
    c. sound source, transducer, amplifier, transport system
    d. sound source, microphone, record button, playback

16. Weak batteries slow down the transport system resulting in:
    a. fast playback
    b. no playback
    c. distorted playback
    d. slow playback

17. To temporarily stop a recorder during a tape session, depress:
    a. record control
    b. pause control
    c. stop control
    d. all of the above

18. The aux input jack bypasses _____ to produce _____ quality than using the mic jack.
    a. recorder's amplifier — higher
    b. recorder's mic — better
    c. recorder's amplifier — lower
    d. all of the above

Coyote, Tascam 225, and 3 Projectors

1. Audio should be recorded on both tracks and pulses should also be recorded on both.
   a. True
   b. False

2. Control tones will be recorded on either track 1 or track 2, but controls and patch cords must be set for the appropriate track.
   a. True
   b. False

3. To program the Coyote:
   a. positrack control should be on
   b. sequence control should be on 3
   c. depress power control
   d. one should do all of the above

4. To cause projector trays to return home to the beginning of the program:
   a. manually rotate them
   b. program a home cue
   c. program a wait command
   d. perform all of the above

5. Parts of a Coyote program should be saved as they are completed because:
   a. a power failure could cause a loss
   b. audiotape is a convenient storage medium
   c. it is simple to determine where you stopped last time
   d. all of the above

6. A program saved on audiocassette from the Coyote can be restored if there is a power failure.
   a. True
   b. False

7. When running a finished program, all projectors should be set on:
   a. off
   b. on
   c. fan

8. The memory of the Coyote may be cleared by:
   a. depressing reset and clear at the same time.
   b. turning the power off on the Coyote
   c. someone tripping over the power cord and jerking it out of the wall outlet
   d. any of the above

9. To get out of the memory verify mode, depress:
   a. reset
   b. reset and clear
   c. step/escape (step/esc)
   d. home

10. Before programming the Coyote, one should depress:
    a. memory load (mem load)
    b. reset
    c. reset and clear
    d. home

11. It is necessary to set the projectors to lamp in order to:
    a. focus them
    b. check size and alignment
    c. check border alignment on slides
    d. perform all of the above

12. The cues are automatically recorded on cassette tape as the Coyote is programmed.
    a. True
    b. False

13. To play back a program, it is important to:
    a. depress reset and clear simultaneously
    b. change patch cords
    c. reset trays to number 1 slot
    d. do all of the above

14. The track 1 input pan should be turned to the far left, and the track 2 input pan should be turned to the far right.
    a. True
    b. False

17. The VHS camcorder is a portable and convenient video tool because:
    a. it is a camera with a built in VCR
    b. it features auto-focus and auto-white balance
    c. it uses standard 1/2-inch VHS videocassette
    d. all of the above

18. On the VHS camcorder the battery status indication in the viewfinder reads E----F when:
    a. the battery is 1/2 charged
    b. the battery is fully charged
    c. the battery is fully discharged
    d. the battery needs replacing

19. When shooting on location with the VHS camcorder, what should be done during every pause in shooting to save battery power?
    a. push rec review function
    b. push standby button
    c. push high speed shutter function
    d. set edit switch to on

20. CBS television began experimenting during the early 1970's with portable video cameras and videotape recorders for news. This technology was labeled:
    a. newsfilm
    b. electronic field production (EFP)
    c. eyewitness news
    d. electronic newsgathering (ENG)

Slot-load 16mm Sound Motion Picture Projector

1. The automatic threading mechanism on an Eiki slot-load is set by:
   a. rotating the inching knob control
   b. turning the function control to stop
   c. rotating the focusing control
   d. doing none of the above

2. The inching knob control is used to:
   a. measure the length of the film to be shown
   b. advance the film manually to determine if it has been threaded correctly
   c. elevate the projector
   d. do none of the above

3. To reverse film back to beginning of vital content:
   a. move function control clockwise to off or stop position and then to reverse position
   b. move function control clockwise to sound and then to reverse position
   c. move function control counterclockwise to off or stop and then to reverse

4. A slot-load projector accepts the film:
   a. lengthwise
   b. at its head end
   c. sprocket holes toward the operator
   d. a and c above
   e. in none of the above configurations

5. To rewind the film, one should:
   a. switch the feed and take-up reels
   b. lower the reel arms
   c. leave the feed and take-up reels where they are
   d. a and b above

Manual-thread 16mm Sound Motion Picture Projector

1. The take-up reel of a motion picture projector should be at least as large as the feed reel.
   a. True
   b. False

2. Most 16mm sound motion picture projectors project films with optical soundtracks.
   a. True
   b. False

3. To correct a motion picture image that starts to flicker or jump:
   a. reform the loop with loop restorer
   b. turn off the amplifier
   c. replace the exciter lamp
   d. turn off the lamp and motor

4. The function of the photoelectric cell in a motion picture projector is to:
   a. act as a light source for illuminating the soundtrack
   b. act as a light source to project the picture
   c. convert light variations on the soundtrack into electric impulses

5. When the sound and picture are not in sync, it is usually because the:
   a. gate is not closed
   b. lower loop is not the correct size
   c. projector is not set on the correct speed
   d. exciter lamp is weak
   e. framing device is out of adjustment

6. The most desirable order of turning off the switches after showing a motion picture is:
   a. projection lamp, motor switch, volume control
   b. motor switch, volume control, projection lamp
   c. volume control, projection lamp, motor switch
   d. motor switch, projection lamp, volume control

7. If the sound from a motion picture is gutteral (too low in pitch), the probable cause is that the:
   a. speaker plug is only partially inserted in the output jack
   b. amplifier is partially warmed up
   c. amplifier is not turned on
   d. projector is set on "silent" speed

8. The exciter lamp indicator on a 16mm projector indicates whether the:
   a. amplifier is on
   b. film is threaded correctly
   c. projection bulb is functional
   d. machine is warmed up

9. If the film breaks during a classroom showing, you should:
   a. postpone the showing until the next day or until the film is repaired
   b. fasten the ends together with a paper clip or cellophane tape and continue the showing
   c. overlap the film on the take-up reel and continue the showing

10. If the sound from the speaker of a 16mm motion picture projector is garbled, what is the most probable cause?
    a. loss of the lower loop
    b. the film is not tight around the sound drum
    c. the exciter lamp is turned out
    d. slippage in the take-up reel

11. Where is the sound in relation to the picture on 16mm motion picture film?
    a. exactly beside the picture
    b. 26 frames ahead of the picture
    c. 26 frames behind the picture
    d. it depends on the type of soundtrack

12. Which one of the following can give the most realistic reproduction of reality on the screen?
    a. overhead projector
    b. opaque projector
    c. slide projector
    d. 16mm sound motion picture projector

Videotape Systems

1. A high-pitched squeal (feedback) emitted by the monitor/TV during recording is probably caused by:
   a. holding the mike too far away from the recorder
   b. depressing the record button without depressing the play button
   c. leaving the monitor volume control turned on
   d. pointing the camera at the mike while both are on

2. A traditional studio production facility consists of a number of video sources that are all connected to a:
   a. character generator
   b. audio mixer
   c. switcher
   d. film chain

3. A video camera should never be pointed at:
   a. the monitor
   b. the sun
   c. a mirror
   d. the floor

4. The video camera is connected to the recorder by:
   a. audio and video cables
   b. a video cable
   c. an audio cable
   d. a power cable

5. A video camera should be focused while:
   a. the zoom/focus control is in the extreme close-up mode
   b. the zoom/focus control is in the extreme wide-angle mode
   c. the aperture control is wide open
   d. the aperture control is shut down

6. To check the audio and video levels before making a videotape, you should depress the:
   a. fwd and record buttons at the same time
   b. play button
   c. record button
   d. audio limiter button

7. The control for adjustin[g] videotape is located on:
   a. the VCR
   b. the camera

8. What should always be [done before] production?
   a. playback a different [tape] and check the audio
   b. make a test recordin[g]
   c. put a label on the ta[pe] program
   d. adjust the color, brig[htness]

9. You have just finished [recording, which of the] following should you do[?]
   a. unplug the camera fr[om]
   b. check the audio leve[l]
   c. replace the lens cap
   d. play the tape back

10. When playing back a vi[deo on channel] 2, the audio monitor swi[tch should be in the]
    a. channel 1 position
    b. mix position

11. Electronic field producti[on is] similiar to filmmaking bec[ause]
    a. film is used instead o[f tape]
    b. the footage is edited
    c. the footage is edited
    d. multiple cameras are

12. When playing back a tap[e you hear sound] but cannot see a picture[, probable cause is]
    a. wrong type of micro[phone]
    b. color tape played on
    c. broken video cable f[rom]
    d. audio output connec[tion]

Microcomputers

1. Computers are defined as information processors.
   a. True
   b. False

2. Hard copy refers to computer information printed to paper.
   a. True
   b. False

3. Microcomputers are single-purpose machines; internal instructions cannot be changed to perform different tasks.
   a. True
   b. False

4. Documentation refers to the detailed instructions that control the entire computer system.
   a. True
   b. False

5. Random access memory controls all of the computer's internal calculations and decision-making routines as instructed by the software.
   a. True
   b. False

6. A diskette that is described as 360K has the capability of storing approximately 360,000 characters.
   a. True
   b. False

7. Information in RAM is not lost when the computer is turned off.
   a. True
   b. False

8. It is important to save your work often to nonvolatile media such as diskettes.
   a. True
   b. False

9. Disk drives are input devices; they can only recieve information from the central processing unit.
   a. True
   b. False

10. Word processors are mainly used for creating graphics such as pictures or charts.
    a. True
    b. False

11. Computerized drills are used to teach new skills or concepts to the student.
    a. True
    b. False

Photographic Copystand

1. The best lens for use on a copystand is:
   a. regular 50mm lens
   b. macro or micro lens
   c. close-up lenses
   d. extension tubes for a 50mm regular lens

2. Focusing a camera mounted on a copystand is easier than with the hand-held technique.
   a. True
   b. False

3. The hot spot from each flood lamp should land:
   a. on the exact center of the material to be photographed
   b. on a spot just beyond the material to be photographed
   c. on the camera
   d. on the wall behind the camera for diffusion of light

4. It is necessary to set the ASA adjustment on the camera:
   a. prior to taking each picture
   b. after taking each picture
   c. when a roll of film is loaded into the camera

5. As the camera is raised on the copystand, the image in the viewfinder:
   a. gets larger
   b. gets smaller
   c. does not change
   d. stays in focus

6. You know that the batteries are working if:
   a. the meter needle moves
   b. a red light glows
   c. numbers appear in a window
   d. any of the above, depending on the camera

7. If the meter needle does not move, you should:
   a. turn on the meter switch
   b. move the film advance lever to the stand-off position
   c. lightly touch the shutter release button
   d. any of the above, depending on the camera

8. The aperture control or f-stop regulates the:
   a. time the shutter is left open
   b. amount of light that passes through the lens to the film
   c. both a and b
   d. neither a nor b

9. Shutter settings must be made only on the numbered intervals.
   a. True
   b. False

10. Correct exposure on the Pentax camera used in this assignment is achieved when:
    a. the needle is at the bottom of the notch in the viewfinder
    b. the needle is at the center of the notch in the viewfinder
    c. the needle is at the top of the notch in the viewfinder
    d. none of the above conditions is present

11. Exposure settings for work on the copystand should always be determined by:
    a. using a white card
    b. using a gray card
    c. using a black card
    d. exposing for the subject

12. The ASA adjustment on the camera is necessary because films vary in sensitivity to:
    a. humidity
    b. temperature
    c. light

13. The camera lens is focused by:
    a. pushing the shutter release
    b. turning the barrel of the lens
    c. moving the camera up or down the support post

14. It has been suggested that the overhead room lights be turned off because:
    a. the lighting would be too flat
    b. unwanted reflections would be minimized
    c. the extra energy would raise the room temperature level too high

Lettering

1. The gaps left by the cardboard stencil should be filled in to make a more legible and attractive letter.
   a. True
   b. False

2. Circular letters such as C, D, and O must be placed close to neighboring letters to equalize the white space between letters.
   a. True
   b. False

3. Legibility of lettering is controlled primarily by:
   a. style and size
   b. style and spacing
   c. size and contrast
   d. spacing and contrast

4. Lettering on overhead transparencies should be at least:
   a. 1/4 inch high
   b. 1/2 inch high
   c. 3/4 inch high
   d. 1 inch high

5. The most easily read letters are
   a. slanted to the right with serifs
   b. slanted to the left without serifs
   c. straight, simple, and without serifs
   d. straight, simple, and with serifs

6. Optical spacing:
   a. equalizes spaces between letters
   b. equalizes distance between letters
   c. provides a rigid box within which each letter is placed
   d. b and c

7. Dry-transfer letters are also known as:
   a. press-on letters
   b. stencils
   c. rub-on letters
   d. transfer type
   e. a, c, and d

Thermal Copying

1. The amount of exposure to infrared rays in a thermal copier is regulated by adjusting the dial to vary the:
    a. speed with which the original passes under the lamp
    b. intensity of the light source
    c. number of backing sheets used
    d. temperature in the copier

2. The thermal process will reproduce only those configurations which:
    a. will absorb heat
    b. reflect light
    c. do not absorb heat
    d. a and b
    e. b and c

3. The original copy should always be inserted into the thermal copying machine in which position?
    a. facedown position
    b. faceup position
    c. it depends on the thermal material you select
    d. it does not matter

4. Normally the thermal process will copy:
    a. xerox
    b. ditto
    c. ball point and felt pens
    d. India ink
    e. a and d

5. Thermal transparencies should be inserted into the machine with the notched or cut corner positioned in the:
    a. upper left
    b. lower right
    c. upper right
    d. direction preferred by the user

6. Print-through of images from the other side of an original will sometimes be reduced by using:
    a. a sheet of plain white paper under the original
    b. a dark-colored backing sheet under the original
    c. a transparent carrier for the original
    d. thicker padding
    e. any of the above

Spirit Duplication

1. Extraneous dye along the edges of a spirit master may be readily removed by:
   a. cutting away the dye-stained margins
   b. dissolving it with duplicator fluid
   c. covering it with cellophane tape
   d. a and c
   e. a, b, and c

2. Multiple colors may be added on the spirit master by:
   a. using alternate dye sheets
   b. using special felt-tip pens
   c. changing the ink in the duplicator
   d. additional runs through the thermal copier

3. The spirit duplication process involves:
   a. the transfer of aniline dye from the back side of a master sheet
   b. forcing ink through openings in a wax-coated tissue
   c. light sensitive paper
   d. none of the above

4. The spirit duplication process has which of the following limitations?
   a. errors are difficult to correct
   b. masters are very expensive
   c. equipment is rarely available in schools
   d. masters wear out after a few hundred copies

5. After a spirit master has been prepared, selected areas may be eliminated by:
   a. scraping off the dye with a sharp blade
   b. covering the unwanted image with tape
   c. cutting out a section of the master (in some cases)
   d. a and b
   e. all of the above

Materials Preservation

1. Materials to be placed in a dry-mount press are placed in a protective envelope of clean paper to:
   a. equalize the heat over the surfaces to be mounted or laminated
   b. prevent adhesive from contacting the heating element or cushion pad
   c. protect the material from adhesive and dirt that might be on the platen
   d. b and c

2. A _____ is used to secure materials in position before inserting them in a dry mount press.
   a. cast-iron weight
   b. wet sponge
   c. tacking iron
   d. burnishing tool

3. Dry-mounting tissue is a thin sheet of paper coated with a heat-sensitive adhesive on:
   a. one side.
   b. both sides.

4. Dry-mounting may be done easily and quickly with a:
   a. dry mount press
   b. household iron
   c. combination of rubber cement and 50 lbs of pressure.
   d. a and b

5. Laminating film is a tough, mylar material that has been coated with a heat-sensitive adhesive on:
   a. one side
   b. both sides

6. Upon application of high temperatures and pressure, the heat-sensitive adhesive on laminating film:
   a. causes the film to turn blue.
   b. softens and adheres to most clean surfaces.
   c. both a and b.

7. It would be unwise to laminate a color photograph or other plastic-coated visual because the high temperature required would also melt the plastic on the visual.
   a. True
   b. False

8. The formation of bubbles under laminated surfaces can be prevented by _____ unmounted materials.
   a. smoothing
   b. preheating
   c. tacking

9. Materials that will be exposed to moisture should be laminated on front and back and have a narrow border to seal them.
   a. True
   b. False

The Copyright Law and Its Implications

1. The most recent United States copyright law (U.S. Public Law 94-553) took effect on January 1:
   a. 1955
   b. 1976
   c. 1978
   d. 1987

2. There is no universal international copyright agreement.
   a. True
   b. False

3. While the 1978 law does not require registration of copyright, registration is necessary before any infringement suits may be filed in court.
   a. True
   b. False

4. A work may be registered at any time within the life of the copyright.
   a. True
   b. False

5. Under the current law, copyrighted works receive protection during the author's lifetime plus:
   a. 20 years
   b. 50 years
   c. 100 years
   d. 150 years

6. When a copyright term runs out, the material is considered to be in _____ and may be freely copied and/or performed.
   a. public interest
   b. public domain
   c. expired authorship

7. Fair use permits limited copying of material without permission from, or payment to, the copyright owner.
   a. True
   b. False

8. To be considered fair use, a particular instance generally must pass the tests of:
   a. nonprofit educational use
   b. nature of copyrighted work
   c. amount of the work used in relation to the whole
   d. effect of the use on potential market for the work
   e. all of the above

9. One should avoid copying slides, filmstrips, prerecorded video, or film if the portion copied includes the creative essence of the work.
   a. True
   b. False

10. Nonprofit educational institutions may record off-air TV broadcasts (including cable) and retain a copy for a period up to 45 consecutive calendar days, after which time it must be erased.
    a. True
    b. False

11. Congress amended the Copyright Act to include a specific section on computer software in:
    a. 1970
    b. 1979
    c. 1980
    d. 1986

Study Question Answer Sheet

Projection (General)

1. a
2. a
3. a
4. b
5. b
6. a
7. a
8. a
9. d
10. b
11. c
12. d
13. a
14. b
15. a
16. a
17. d
18. e
19. a
20. e

Multiple-load Slide Projector

1. a
2. a
3. a
4. b
5. a
6. c

Opaque Projector

1. a
2. b
3. c
4. c
5. e
6. b
7. c

Overhead Projector

1. b
2. d
3. d
4. d
5. d

Audio (General)

1. b
2. b
3. a
4. c
5. b
6. d
7. a
8. b
9. b
10. a
11. c
12. d

Filmstrip Projector

1. b
2. c
3. c
4. d
5. c
6. b
7. c
8. a
9. c
10. d

Coyote, Tascam 225, and 3 Projectors

1. b
2. a
3. d
4. b
5. d
6. a
7. c
8. d
9. c
10. c
11. d
12. b
13. d
14. b

Photographic Copystand

1. b
2. a
3. b
4. c
5. b
6. d
7. d
8. b
9. a
10. b
11. b
12. c
13. b
14. b

Audiocassette Recorder

1. a
2. a
3. a
4. a
5. c
6. c
7. c
8. c
9. e
10. c
11. b
12. d
13. a
14. a
15. c
16. a
17. b
18. a

Slot-load 16mm Sound Projectors

1. b
2. b
3. c
4. d
5. c

Manual-thread 16mm Sound Projectors

1. a
2. a
3. a
4. c
5. b
6. c
7. d
8. a
9. c
10. b
11. b
12. d

Lettering

1. a
2. a
3. b
4. a
5. c
6. a
7. e

The Copyright Law and Its Implications

1. c
2. a
3. a
4. a
5. b
6. b
7. a
8. e
9. a
10. a
11. c

Thermal Copying

1. a
2. a
3. b
4. e
5. c
6. b

Videotape Systems

1. c
2. c
3. b
4. b
5. a
6. c
7. a
8. b
9. c
10. d
11. b
12. c
13. b
14. c
15. b
16. d
17. d
18. b
19. b
20. d

Microcomputers

1. a
2. a
3. b
4. b
5. b
6. a
7. b
8. a
9. b
10. b
11. b

Spirit Duplication

1. d
2. a
3. a
4. d
5. e

Materials Preservation

1. d
2. c
3. b
4. d
5. a
6. b
7. a
8. b
9. a

Materials Production Assignment Sheets

Grading Scale

(What the numbers mean)

10 – Superior
 9 – Excellent
 8 – Very good
 7 – Good enough for classroom use
 6 – Barely adequate for use
 5 – Usable, but you should be embarrassed
 4 – Not recommended for classroom use
 3 – Wasted effort (You may have learned what to do next time, though)
 2-0 – You're not serious

Assignment Sheet 1   Dry-transfer Lettering

Purpose

To provide experience using dry-transfer letters to produce graphic materials with professional-looking lettering.

Objective

Observable Behavior

Letter a nonsense word using all of the letters from your last name on one line and your I.D. number on the second line. Or do other assignment specified by your instructor.

Conditions

Given this book, the necessary equipment and materials, and adequate time to practice.

Level of Acceptable Performance

Lettering may be of any style. It must be optically centered horizontally and vertically. Select a letter size small enough that the finished project will not exceed 11 X 14 inches. Words must be correctly spelled. Letter spacing must be optically satisfying to the evaluator. The panel must be free from visible smears.

Materials and Equipment

Pencil and straightedge
Sheet of dry-transfer letters, 1/4" (18 point) minimum,
　　　　　　　　　　　　　　　　3/4" (54 point) maximum size
Cardboard, 11" X 14"
Dull pencil, ball point pen, or other burnishing tool

Assignment

1. Follow the instructions provided for Dry-Transfer and Lettering Principles.
2. Letter the words as directed under objective.
3. Tape this page to the back of your project and submit it for evaluation.

   Rejected — Word Misspelled

0 1 2 3 4	alignment of letters
0 1 2 3 4	letter spacing (optical)
0 1 2 3 4	position on page
0 1 2 3 4	cleanliness
0 1 2 3 4	freedom from cracks and chips

   2)　　　　　　(= score

Assignment Sheet 2   Cardboard Stencil Lettering

Purpose

This assignment will provide experience with an inexpensive, easy to use, and readily available type of lettering device, which is especially good for letters that are 3/4" to 4" high.

Objective

Observable Behavior

Letter a nonsense word using all of the letters from your last name on one line and your I.D. number on the second line. Or do other assignment specified by your instructor.

Conditions

Given this book, the necessary equipment and materials, and adequate time to practice.

Level of Acceptable Performance

Lettering may be of any size but must be centered horizontally and vertically. Select a letter size small enough that the finished project will not exceed 11 × 14 inches. Words must be correctly spelled. The work should be free from visible smears. Mechanical spacing is acceptable.

Materials and Equipment

Cardboard stencil
Paper or cardboard, 11 × 14 inches maximum
Pencil and ruler

Assignment

1. Follow the instructions provided for Cardboard Stencil and Lettering Principles.

2. Letter the words as directed under objective.

3. Tape this page to the back of the project and submit it for evaluation.

Rejected — Word Mispelled

0 1 2 3 4	alignment of letters
0 1 2 3 4	letter spacing
0 1 2 3 4	position on page
0 1 2 3 4	cleanliness
0 1 2 3 4	consistency of letters (density if filled)
_____ 2)	(= score

Assignment Sheet 3 Thermal Transparency

Purpose

The best way to learn a process is to try it. In this assignment you will produce a transparency on the Thermofax machine.

Objective

Observable Behavior

Produce a transparency from a composite original.

Conditions

Given this book and the necessary equipment and materials.

Level of Acceptable Performance

Submit the original and one properly exposed transparency; i.e., the best copy possible from the original.

Materials

Sheet of 8 1/2" X 11" white paper
Thermal transparency film (direct process single-sheet type, any color)
Thermal copier
No. 2 pencil
Ballpoint pen
Felt tip- or nylon-tip pen
India ink and pen
Tearsheet

Assignment

1. Select a tearsheet from a magazine that has some black and some colored images.

2. Cut this tearsheet to a size that will cover no more than half of the 8 1/2" X 11" paper.

3. On the lower half of the paper, letter
   "NUMBER TWO PENCIL"
   "BALLPOINT PEN"
   "FELT-TIP PEN"
   "INDIA INK"
   with the appropriate instruments as shown. Print your I.D. number along the lower left edge with no. 2 pencil.

4. Position the small tearsheet above the lettering on the 8 1/2" X 11" page. You may fasten it with transparent tape if you wish.

5. Proceed as explained for Thermal Copying.

6. Tape transparency and this page to the composite. On the back of the composite, write an explanation of why some of the original copied well and some did not.

7. Submit all 3 items for evaluation.

0 1 2 3 4	clear borders
0 1 2 3 4	ink area blackness
0 1 2 3 4	absence of copy-through
0 1 2 3 4	blur (over exposure)
0 1 2 3 4	dropout (under exposure)
2)	(= score

Assignment Sheet 4 Thermal Spirit Master

Purpose

Frequently one would like to duplicate elaborate drawings or complete pages of print for class distribution. The thermal spirit master provides an economical and quick way of accomplishing this time-consuming, or sometimes impossible, task.

Objective

Observable Behavior

Produce a replica of a tearsheet by the spirit master process.

Conditions

Given this book and the necessary equipment and materials.

Level of Acceptable Performance

Submit the original, master, and two bright, clear, purple copies of a selected article, advertisement, or cartoon.

Materials and Equipment

Thermal copier
Plastic carrier for thermal copier
Spirit duplicator with fluid
Several sheets of duplicator paper
Thermal spirit master
Original

Assignment

1. Select an article, advertisment, or cartoon that you wish to duplicate.

2. Follow the steps explained for Thermal Spirit Master Preparation.

3. Run a few copies of the completed master as explained in the unit on Spirit Duplication.

4. Print your I.D. number on the back of one of the copies. Staple the master and the original between that and another copy and submit for evaluation.

0 1 2 3 4	contrast and clarity
0 1 2 3 4	absence of copy-through
0 1 2 3 4	evenness of copy
0 1 2 3 4	absence of wrinkles
0 1 2 3 4	cleanliness of borders
2)	(= score

Assignment Sheet 5   Direct Spirit Master

Purpose

The spirit duplication process is a most valuable technique which teachers use almost daily. This exercise is to provide experience in producing a master and running copies on the duplicator.

Objective

Observable Behavior

Produce a two-color spirit master by drawing and typing. Produce two copies from your master to be submitted for evaluation.

Conditions

Given this book and the necessary equipment and materials.

Level of Acceptable Performance

Submit two crisp, unwrinkled copies and your original master. Corrections should not be visible on the copies.

Materials and Equipment

2 spirit masters (1 purple, plus 1 red, blue, or green)
Typewriter and ball point pen
Spirit duplicator with fluid
Several sheets of clean 8 1/2 × 11 inch duplicator paper

Assignment

1. Follow the steps listed for Direct Spirit Master Preparation. Type the following quotation (or several lines of your choice) using the purple spirit master dye sheet. "The more extensive a man's knowledge of what has been done, the greater will be his power of knowing what to do." Disraeli

2. Type the following quotation (or several lines of your choice) using a second color. "A teacher affects eternity; he can never tell where his influence stops." Henry Adams

3. Print your I.D. number in one of the corners of the spirit master. This may be either of the two colors.

4. Follow the steps listed for Spirit Duplication to produce multiple copies of your master.

5. Staple the master between two copies, attach this page, and submit for evaluation.

Rejected — Second Color

0 1 2 3 4 5	contrast and clarity
0 1 2 3 4 5	spelling and strikeovers
0 1 2 3 4 5	evenness of copy
0 1 2 3 4 5	wrinkles
2)	(= total

Assignment Sheet 6 Dry Mounting

Purpose

If a picture or other graphic is worth saving, it is worth preserving! Tearsheets (pages from magazines, etc.) quickly become dog-eared, ripped, and soiled if not treated for permanence in some manner. The purpose of this assignment is to provide experience in preserving tearsheets by dry mounting.

Objective

Observable Behavior

Mount a visual to backing material.

Conditions

Given this book and the necessary equipment and materials.

Level of Acceptable Performance

Produce a mounted visual that is completely adhered; free of wrinkles, bubbles and creases; and has aesthetically pleasing borders (i.e., side borders equal, bottom border wider than top).

Materials and Equipment

Dry-mount press and tacking iron or household iron
Paper cutter or sharp blade
Tearsheet
Cardboard backing material, 11" × 14" maximum
MT 5 dry-mounting tissue (or other brand)

Assignment

1. Follow steps listed for Dry Mounting.

2. Centering the visual on the backing materials may be more easily accomplished if it is moved to a position flush with one edge of the backing material. A measurement of the distance from the opposite edge of the backing material to the visual may be halved to determine the correct position for centering. Always leave more space at the bottom of the completed mounting than at the top, for aesthetic reasons.

3. Print your I.D. number on back of the project in ink.

4. Attach this page to the project and submit for evaluation.

0 1 2 3 4	adherence
0 1 2 3 4	cleanliness
0 1 2 3 4	creases and surface blemishes
0 1 2 3 4	attention to edges
0 1 2 3 4	placement
2)	(= score

Assignment Sheet 7 Laminating (With the Dry-mount Press)

Purpose

Laminating film may be used to protect and preserve a variety of materials from spills, fingerprints, etc. The film is tough material that will take a great deal of stress. This assignment will give the student an introductory experience with the laminating process.

Objective

Observable Behavior

Mount and laminate a visual that may be used as a study print.

Conditions

Given this book and the necessary equipment and supplies.

Level of Acceptable Performance

Produce a laminated print that is free from bubbles and wrinkles and has aesthetically pleasing borders.

Materials and Equipment

Dry-mount press and tacking iron or household iron
Paper cutter or sharp blade
Selected visual
Thermal laminating film
Visual mounted on cardboard, 11" × 14" maximum

Assignment

1. Follow steps listed for Dry Mounting, i.e., mount selected visual on backing material before proceeding to step 2.
2. Follow steps for Laminating.
3. Print your I.D. number on back of project in ink.
4. Attach this page to the project and submit for evaluation.

Rejected — Dry mount First

0 1 2 3 4	adherence
0 1 2	cleanliness
0 1 2 3 4	blisters and bubbles
0 1 2 3 4	creases and surface blemishes
0 1 2 3 4	attention to edges
0 1 2	placement

2) (= score

Instructional Equipment Checkout Sheet

Overhead Projector 2 minutes

Connect power cord
Turn projector on
Position material on stage
Adjust image to screen size
Focus image
Indicate head assembly
Indicate transport handle
Restore to storage conformation
8 Checkpoints

Opaque Projector 3 minutes

Connect power cord
Raise lens cover
Engage platen release lock
Raise platen release lever
Position material for projection
Indicate position of heat-absorbing glass
Lower projector onto material
Turn lamp on
Position projected image on screen
Focus image
Adjust image elevation
Activate pointer
Restore to storage conformation
13 Checkpoints

Filmstrip Projector with Audiocassette Playback 5 minutes

Open case and position projector
Connect power cord
Adjust image to screen size
Thread filmstrip
Focus image
Adjust framer
Adjust film speed control
Adjust for audible or inaudible pulse
Load audiocassette
Show program
Rewind tape and remove from machine
Rewind filmstrip and return to storage container
Restore projector to storage conformation
13 Checkpoints

Multiple-load Slide Projector 5 minutes

Connect power cord
Plug in remote control cord
Load and install slide tray
Insert lens
Set automatic timer at m
Move function control to low or high
Prefocus to adjust image to screen size
Fine-focus screen image with first slide
Project slides
Remove slide tray
Move function control to off
Remove slides from slide tray
Restore to storage conformation
13 Checkpoints

Portable Screen 3 minutes

Extend tripod legs
Attach screen to center post
Extend center post
Readjust center post, extend screen to maximum
Lower screen into housing
Restore to storage conformation
6 Checkpoints

Record Player 4 minutes

Open case and connect power cord
Turn amplifier on
Turn turntable on
Set turntable speed control
Adjust tempo
Place record on turntable
Position pickup needle on record
Adjust volume
Adjust tone
Indicate output jack
Indicate input jack
Turn machine off
Restore to storage conformation
13 Checkpoints

Sharp RD-671AV Audiocassette Tape Recorder
5 minutes

Connect power cord
Check for spindle rotation
Stop recorder
Insert cassette tape
Connect microphone
Make a recording
Rewind tape
Play tape
Record sync pulse track
Rewind tape
Play tape with sync signals
Sync record indicator lamp will light
Rewind tape
Restore to storage conformation
14 Checkpoints

Wollensak 2870 and Kodak Ektagraphic III
10 minutes

Set up slide projector
Arrange slides
Place tray on projector
Connect power cord to outlet
Remove cover on Wollensak
Connect power cord to outlet
Connect Wollensak to Ektagraphic III with patch cord
Turn on power to Wollensak and Ektagraphic
Insert audiocassette in Wollensak
Rewind cassette to beginning of program
Engage sync mode record and start on Wollensak
Record a series of sync pulses
Stop recording
Rewind tape
Return slide tray to 0 or beginning

Play back synchronized program
Adjust volume
Stop tape
Rewind tape
Rotate slide tray to 0
Eject tape
Turn off recorder and projector
Disconnect patch cord and power cords
Replace Wollensak cover
24 Checkpoints

AVL Coyote, Tascam 225 Syncaset, and AVL Synchronizer
20 minutes

Connect power cord
Check for spindle rotation
Connect track 1 of tape recorder to amplifier
Separate signals of two tracks
Insert cassette tape
Switch patch cord to listen to track 2
Erase either track 1 or 2
Rewind tape
Program Coyote
Connect Coyote to audio deck
Separate signals
Adjust volume of control signals
Insert cassette tape
Start recording
Stop recording
Rewind tape
Verify Coyote program
Load Coyote from tape
Switch patch cords
Prepare Coyote and Tascam 225 to record combined program
Record program
Rewind and play program
22 Checkpoints

Slot-load 16mm Sound Motion Picture Projector
6 minutes

Position projector
Remove cover
Position reel arms
Place take-up reel on take-up reel spindle
Place supply reel on supply reel spindle
Turn motor and lamp on
Adjust projected light pattern
Turn motor and lamp off
Set automatic threading mechanisms
Thread film
Attach leader to take-up reel
Start projector
Refocus
Adjust volume and tone controls
Turn projector off
Reverse film back to beginning
Turn volume to low
Turn lamp on and increase volume slowly
Adjust framer
Stop film and back up to repeat a scene
At conclusion of film, gradually turn down volume
Turn motor off after trailer has run through machine
22 Checkpoints

Manual-thread 16mm Sound Motion Picture Projector
6 minutes

Open case and connect power cord
Attach speaker plug
Extend film supply and take-up reel arms
Attach drive belt
Turn amplifier on
Prefocus to adjust image to screen size
Thread film

Rotate manual control knob to observe whether film is correctly threaded
Move function control to lamp
Focus image
Adjust volume
Adjust fidelity
Adjust tone
Adjust elevation
Adjust framing
Indicate upper and lower sprockets
Fade out sound and turn lamp off
Turn projector off
Rewind film
Restore to storage conformation
20 Checkpoints

Panasonic 1/2" VHS Camcorder Record/Playback Unit *10 minutes*

Connect power supply
Turn power on
Set white balance control
Remove lens cap
Insert cassette
Locate general operation controls
Locate tape running controls
Rewind tape
Extend electronic viewfinder
Set focus mode selector control
Identify macro close-up capability
Record
Rewind
Playback
Restore to storage conformation
15 Checkpoints

3/4" Videocassette Recorder/Player *15 minutes*

Connect power cord
Turn power on
Load cassette
Rewind cassette, if necessary
Connect microphone
Adjust audio level control
Set input control
Connect camera to recorder
Connect camera to power
Remove lens cap
Set white balance
Connect monitor to power
Turn monitor power control on
Select TV/VTR/LINE control on monitor
Make a recording
Stop recording
Rewind videocassette
Play videocassette
Adjust tracking and skew control, if necessary
Adjust picture and sound controls on monitor
20 Checkpoints

NOTES

NOTES

NOTES

NOTES

NOTES